Cambridge Tracts in Mathematics
and Mathematical Physics

No. 60

MINIQUATERNION GEOMETRY

An introduction to the study of projective planes

MINIQUATERNION GEOMETRY

AN INTRODUCTION TO THE STUDY OF PROJECTIVE PLANES

T. G. ROOM, F.R.S.

AND

P. B. KIRKPATRICK, PH.D.

University of Sydney

CAMBRIDGE

AT THE UNIVERSITY PRESS

1971

Published by the Syndics of the Cambridge University Press
Bentley House, 200 Euston Road, London N.W.1
American Branch: 32 East 57th Street, New York, N.Y.10022

Library of Congress Catalogue Card Number: 79–123347

ISBN: 0 521 07926 8

Printed in Great Britain
at the University Printing House, Cambridge
(Brooke Crutchley, University Printer)

CONTENTS

PREFACE

In the study of the elementary properties of fields, one of the difficulties, experienced perhaps more by the teacher than the taught, is that there are very few examples of explicit simple systems which resemble fields in having two group operations but do not satisfy all the conditions imposed on a field. 'Miniquaternion algebra' provides one of the simplest such systems: there are only nine elements, and the addition and multiplication tables are easy to remember, although multiplication is neither commutative nor left-distributive over addition.

From the miniquaternion algebra three different plane geometrical systems can be constructed, and the field of order nine gives rise to a fourth geometry. While these four geometries are the main subject of this book, many of the ideas developed are of much more general significance. This book is intended to provide an introduction to the relatively new subject of Projective Planes.

The authors have assumed only a knowledge of the simpler properties of groups, fields, matrices and transformations (mappings), so that the reader who has completed a first course in abstract algebra should have an adequate background for this book.

It is intended that the exercises, which vary from the very easy to the quite difficult, should form an integral part of the book.

The authors wish to record their thanks to the University of Sydney, which provided the environment in which the idea of writing this book took shape and grew, and to Westfield College in the University of London for the hospitality afforded them in the Michaelmas Term 1969 during which the final draft was written. At this time one of us (P.B.K.) received financial support from the University of Sydney as Eleanor Sophia Wood Travelling Fellow, and the other from Westfield College as a Senior Visiting Fellow.

We wish to thank also Dr J. A. Todd for his many valuable suggestions for the improvement of the text, and the Editors of the C.U.P. who have guided and assisted us through all the stages of publication.

T. G. ROOM

P. B. KIRKPATRICK

London and Sydney
June, 1970

PART I

ALGEBRAIC BACKGROUND

We are to be concerned with geometry in planes in which the coordinates of points and the equations of lines can be expressed in terms of elements either of a 'field' or of an algebraic system which satisfies all but one of the axioms which characterize a field. Familiar examples of fields are the algebraic systems whose elements are the rational numbers or the real numbers or the complex numbers; we are more particularly interested in fields with only finite numbers of elements. The 'Galois field' GF (p) of prime order p (named after E. Galois, 1811–32) has for its elements the numbers $0, 1, \ldots, p-1$ and the rules of composition are those of ordinary arithmetic, except that sums and products are 'reduced modulo p'; that is, every number is replaced by the remainder on dividing it by p.

In all these fields multiplication is commutative, and, in fact, the commutative law for multiplication is often included among the axioms for a field. But in the natural evolution of coordinate systems from *geometric* axioms, systems based on commutative fields appear as special cases of systems based on non-commutative (or 'skew') fields. We shall therefore designate as a 'field' a system which is not necessarily commutative in multiplication.

The first strictly non-commutative field to be devised was the system of 'quaternions', invented in 1843 by Sir William Rowan Hamilton to represent the operation of compounding rotations in ordinary Euclidean space. Quaternion numbers are written as $ai+bj+ck+d$ and are compounded in accordance with the axioms for a field. a, b, c, d are real numbers, and i, j, k are 'units' which are independent under addition, and under multiplication satisfy the conditions:

$$ij = k, \quad jk = i, \quad ki = j,$$
$$ji = -k, \quad kj = -i, \quad ik = -j,$$
$$i^2 = -1, \quad j^2 = -1, \quad k^2 = -1.$$

There is a finite system, having only nine elements, which bears a strong resemblance to Hamilton's quaternions. But for the simplification consequent on using such a small system we pay the price of losing one of the distributive laws. The system is therefore not a field.

A finite commutative field† of order 9 exists also, but its definition is a little more complicated than that of the fields of prime order. This field and the 'miniquaternion system' are constructed in the first chapter. In succeeding chapters we discuss how any field may be used to coordinatize a 'projective plane', and also how, from the miniquaternion system, three quite different projective planes are obtained by using various methods of coordinatization.

† J. M. Wedderburn proved in 1905 that every finite field is commutative. For a complete account of this theorem see, for example, I. N. Herstein, 1964, p. 318.

CHAPTER 1

TWO ALGEBRAIC SYSTEMS WITH NINE ELEMENTS

1.1 Near-fields of order 9

The class of finite fields is included in a larger class of algebraic systems called 'finite near-fields', which we consider here because it includes also the miniquaternion system. In fact the miniquaternion system is the smallest near-field which is not a field (H. Zassenhaus, 1936).

DEFINITION 1.1.1 A *finite near-field* is a system $(\mathscr{S}, +, \circ)$, where $+, \circ$ are binary operations on the set $\mathscr{S}$ and

(1) $\mathscr{S}$ is finite,

(2) $+$ is a commutative group operation on $\mathscr{S}$, with identity 0,

(3) $\circ$ is a group operation on $\mathscr{S}-\{0\}$, with identity 1,

(4) $\circ$ is right-distributive over $+$, that is,

$$(\xi+\eta)\circ\zeta = (\xi\circ\zeta)+(\eta\circ\zeta) \quad \text{for all} \quad \xi, \eta, \zeta \in \mathscr{S}, \dagger$$

(5) $\xi\circ 0 = 0$ for all $\xi \in \mathscr{S}$.

For convenience we make the usual convention that $\circ$ is to take precedence over $+$, so that the right-distributive law may be written more simply

$$(\xi+\eta)\circ\zeta = \xi\circ\zeta+\eta\circ\zeta.$$

If both distributive laws hold in a finite near-field then the system is a finite (commutative) field. We are to construct two near-fields of order 9: one a field, the other the miniquaternion system. In fact these are the only near-fields of order 9 (Zassenhaus, 1936).

THEOREM 1.1.1 *If* $(\mathscr{S}, +, \circ)$ *is a finite near-field then* $0\circ\xi = 0$ *for all* $\xi \in \mathscr{S}$.

† Some authors—for example, H. Zassenhaus—replace the right-distributive law by the left-distributive law.

First

$$(0+0)\circ\xi = 0\circ\xi + 0\circ\xi \quad \text{(right-distributivity)}.$$

But $\qquad 0+0=0 \qquad$ (0 is identity for $+$),

and so $\qquad 0\circ\xi = 0\circ\xi + 0\circ\xi.$

Therefore

$$\begin{aligned} -(0\circ\xi)+0\circ\xi &= -(0\circ\xi)+(0\circ\xi+0\circ\xi) \\ &= [-(0\circ\xi)+0\circ\xi]+0\circ\xi \text{ (associativity of } +) \\ &= 0+0\circ\xi \\ &= 0\circ\xi. \end{aligned}$$

Exercise 1.1.1 Show that in any finite near-field:
(i) $\xi\circ\eta = 0 \Rightarrow$ either $\xi = 0$ or $\eta = 0$,
(ii) $1 \neq 0$ and $-1 \neq 0$.

In any near-field $(\mathscr{S}, +, \circ)$ of order 9, the subset $\mathscr{D} = \{0, 1, -1\}$ of $\mathscr{S}$ plays a special role. For consider the additive subgroup of $\mathscr{S}$ generated by 1. This necessarily contains 0 and -1; $-1 \neq 1$ since $-1 = 1$ would imply $1+1 = 0$, that is, the subgroup has order two, which is impossible since the group $(\mathscr{S}, +)$ has order nine, so that by Lagrange's Theorem any non-trivial subgroup of it has order three or nine. If the subgroup is of order nine then $\mathscr{S}$ itself is generated by 1, that is, the elements of $\mathscr{S}$ are

$$0,\ 1,\ 1+1, \ldots, 1+1+1+1+1+1+1+1,$$

and $$1+1+1+1+1+1+1+1+1 = 0;$$

but

$$(1+1+1)\circ(1+1+1) = 1+1+1+1+1+1+1+1+1 = 0,$$

by repeated application of the right-distributive law, so

$$1+1+1 = 0,$$

giving a contradiction. Therefore the additive subgroup generated by 1 has order three: it must be $\mathscr{D} = \{0, 1, -1\}$, and we must have $1+1+1 = 0$.

$(\mathscr{D}, +, \circ)$ is the Galois field GF(3), which we shall usually denote simply by $\mathscr{D}$.

THEOREM 1.1.2 *In any near-field $(\mathscr{S}, +, \circ)$ of order 9, the subset $\mathscr{D} = \{0, 1, -1\}$ is such that $(\mathscr{D}, +, \circ)$ is the Galois field of order 3.*

EXERCISE 1.1.2 Write out addition and multiplication tables for $\mathscr{D}$.

EXERCISE 1.1.3 Show that in any near-field of order 9

(i) $\xi+\xi+\xi = 0$ for all ξ

(ii) $(-1)\circ\xi = -\xi = \xi\circ(-1)$ for all ξ

(iii) $(-\xi)\circ\eta = -(\xi\circ\eta) = \xi\circ(-\eta)$ for all ξ, η

(iv) $(-\xi)\circ(-\eta) = \xi\circ\eta$ for all ξ, η.

THEOREM 1.1.3 *Suppose $(\mathscr{S}, +, \circ)$ is a near-field of order 9 and $\sigma \in \mathscr{S}$, but $\sigma \notin \mathscr{D}$. Then each element of $\mathscr{S}$ may be written uniquely in the form $a+b\circ\sigma$, with $a, b \in \mathscr{D}$.*

Assume $$a+b\circ\sigma = a'+b'\circ\sigma \quad (a, b, a', b' \in \mathscr{D}).$$

Then $$\begin{aligned} a-a' &= b'\circ\sigma-(b\circ\sigma) \\ &= b'\circ\sigma+(-b)\circ\sigma = [b'+(-b)]\circ\sigma \\ &= (b'-b)\circ\sigma. \end{aligned}$$

If $b' \neq b$, then $\sigma = (b'-b)^{-1}\circ(a-a') \in \mathscr{D}$; but $\sigma \notin \mathscr{D}$, so $b' = b$, which implies that $a' = a$. It follows that there are nine distinct elements $a+b\circ\sigma$ in $\mathscr{S}$ corresponding to the nine pairs a, b. But $\mathscr{S}$ has only nine elements, so the theorem is proved.

This theorem implies that, once an element σ not in $\mathscr{D}$ has been chosen, there is a natural 1–1 correspondence between elements of $\mathscr{S}$ and ordered pairs (a, b) of elements of $\mathscr{D}$:

$$a+b\circ\sigma \leftrightarrow (a, b).$$

If we identify corresponding elements, that is, write

$$(a, b) = a+b\circ\sigma,$$

then since

$$(a_1+b_1\circ\sigma)+(a_2+b_2\circ\sigma) = (a_1+a_2)+(b_1+b_2)\circ\sigma$$

we have $$(a_1, b_1)+(a_2, b_2) = (a_1+a_2, b_1+b_2).$$

This means that the addition operation in $\mathscr{S}$ is uniquely determined by the addition operation of the Galois field $\mathscr{D}$, and thus:

THEOREM 1.1.4 *The additive groups of all near-fields of order* 9 *have the same abstract group structure* (*that is, they are isomorphic*).

1.2 The Galois field $\mathscr{F}$ of order 9

We now construct, using $\mathscr{D}$, a near-field $\mathscr{F}$ of order 9 which, in addition to satisfying the conditions of Definition 1.1.1, is left-distributive, and therefore a field.

The nine elements of $\mathscr{F}$ are the ordered pairs (a, b), $a, b \in \mathscr{D}$. Addition in $\mathscr{F}$ is defined by the rule

$$(a_1, b_1) + (a_2, b_2) = (a_1 + a_2, b_1 + b_2)$$

and multiplication by

$$(a_1, b_1) \times (a_2, b_2) = (a_1 a_2 - b_1 b_2, a_1 b_2 + a_2 b_1),$$

where $a_1 a_2, b_1 b_2$, etc., denote products in $\mathscr{D}$. (Compare with addition and multiplication of ordinary complex numbers when they are written as ordered pairs of real numbers.)

It is easily seen that $+$ is a commutative group operation on $\mathscr{F}$: the commutativity and associativity follow from the same properties of $+$ in $\mathscr{D}$; the identity is $0 = (0, 0)$ and the inverse $-(a, b)$ of (a, b) is $(-a, -b)$.

By the symmetry of the formula for $(a_1, b_1) \times (a_2, b_2)$ and the commutativity of the operations in $\mathscr{D}$, the operation $\times$ is commutative. That $\times$ is a group operation on $\mathscr{F} - \{0\}$ is again straightforward. (The proof is the same as for the complex numbers.) The identity is $1 = (1, 0)$ and the inverse $(a, b)^{-1}$ of $(a, b) \neq (0, 0)$ is $(a/(a^2+b^2), \; -b/(a^2+b^2))$. (If $a, b \in \mathscr{D}$ then $a^2 + b^2 = 0 \Leftrightarrow a = b = 0$.)

Finally, $\times$ is associative:

$$\begin{aligned}
&(a_1, b_1) \times [(a_2, b_2) \times (a_3, b_3)] \\
&\quad = (a_1, b_1) \times (a_2 a_3 - b_2 b_3, a_2 b_3 + b_2 a_3) \\
&\quad = (a_1 a_2 a_3 - a_1 b_2 b_3 - b_1 a_2 b_3 - b_1 b_2 a_3, \\
&\qquad\qquad a_1 a_2 b_3 + a_1 b_2 a_3 + b_1 a_2 a_3 - b_1 b_2 b_3) \\
&\quad = (a_1 a_2 - b_1 b_2, a_1 b_2 + b_1 a_2) \times (a_3, b_3).
\end{aligned}$$

Thus $\times$ is a group operation on $\mathscr{F} - \{0\}$.

The proof of right-distributivity is routine:

$$\begin{aligned}
&[(a_1, b_1) + (a_2, b_2)] \times (a_3, b_3) \\
&\quad = (a_1 + a_2, b_1 + b_2) \times (a_3, b_3) \\
&\quad = (a_1 a_3 + a_2 a_3 - b_1 b_3 - b_2 b_3, a_1 b_3 + a_2 b_3 + a_3 b_1 + a_3 b_2) \\
&\quad = (a_1 a_3 - b_1 b_3, a_1 b_3 + a_3 b_1) + (a_2 a_3 - b_2 b_3, a_2 b_3 + a_3 b_2) \\
&\quad = (a_1, b_1) \times (a_3, b_3) + (a_2, b_2) \times (a_3, b_3).
\end{aligned}$$

Thus $\mathscr{F}$ is a near-field of order 9. But $\times$ is commutative. So $\mathscr{F}$ is also left-distributive; that is, $\mathscr{F}$ is a commutative field.

For simplicity we shall henceforth denote $\times$ by the usual . or juxtaposition.

If we write $a = (a, 0)$ and $\epsilon = (0, 1)$ then we may easily verify that $(a, b) = a + b\epsilon$ and $\epsilon^2 = -1$.

Let us examine the multiplicative group of $\mathscr{F}$. This group is in fact cyclic. It is not generated by ϵ since $\epsilon^2 = -1$, $\epsilon^4 = 1$. But $(1-\epsilon)^2 = 1+\epsilon+\epsilon^2 = \epsilon$, so that $(1-\epsilon)^4 = \epsilon^2 = -1$, and, therefore, $(1-\epsilon)^8 = 1$ and no smaller power of $1-\epsilon$ is equal to 1.

Denote $1-\epsilon$ by ω. The powers of ω are:

$$\begin{aligned}
\omega &= 1-\epsilon, & \omega^5 &= -1+\epsilon, \\
\omega^2 &= \epsilon, & \omega^6 &= -\epsilon, \\
\omega^3 &= 1+\epsilon, & \omega^7 &= -1-\epsilon, \\
\omega^4 &= -1, & \omega^8 &= 1.
\end{aligned}$$

Considering the non-zero elements of $\mathscr{F}$ as powers of ω simplifies multiplication, but makes addition a little more complicated. Addition may be performed without referring back to ϵ by using the relation

$$\omega^2 = -\omega + 1.$$

For example, $\omega^3 = -\omega^2 + \omega = -(-\omega+1) + \omega = -\omega - 1$, so that $\omega^3 + 1 = -\omega$; but

$$\omega^4 = (\omega^2)^2 = (1-\omega)^2 = 1 + \omega + \omega^2 = 1 + \omega + 1 - \omega = -1,$$

and therefore $\omega^3 + 1 = -\omega = (-1).\omega = \omega^4.\omega = \omega^5$.

EXERCISE 1.2.1 From $\omega^2 = -\omega + 1$ deduce the relation $\omega^2 + 1 = \omega^3$.

If $\omega^r = a+b\epsilon$, then $(\omega^r)^3 = a^3+3a^2b\epsilon+3ab^2\epsilon^2+b^3\epsilon^3$. But

$$x \in \mathscr{D} \Rightarrow x^3 = x \quad \text{and} \quad 3x = 0.$$

Also $\epsilon^2 = -1$, so that $\omega^r = a+b\epsilon \Rightarrow w^{3r} = a-b\epsilon$.

Following the practice with complex numbers we use the terminology

DEFINITION 1.2.1 $a+b\epsilon$ and $a-b\epsilon$ are *conjugate* members of $\mathscr{F}$, and we write

$$(a+b\epsilon)^* = a-b\epsilon, \quad \text{that is} \quad (\omega^r)^* = \omega^{3r}.$$

We usually write ω^{r*} for the conjugate of ω^r. Note that

$$\omega^{r*} = \omega^{3r} = (\omega^3)^r = (\omega^*)^r.$$

For future reference we give the complete table for addition, in terms of ω:

Table 1.2.1

$+$	1	ω	ω^2	ω^3	ω^4	ω^5	ω^6	ω^7
1	ω^4	ω^7	ω^3	ω^5	0	ω^2	ω	ω^6
ω	ω^7	ω^5	1	ω^4	ω^6	0	ω^3	ω^2
ω^2	ω^3	1	ω^6	ω	ω^5	ω^7	0	ω^4
ω^3	ω^5	ω^4	ω	ω^7	ω^2	ω^6	1	0
ω^4	0	ω^6	ω^5	ω^2	1	ω^3	ω^7	ω
ω^5	ω^2	0	ω^7	ω^6	ω^3	ω	ω^4	1
ω^6	ω	ω^3	0	1	ω^7	ω^4	ω^2	ω^5
ω^7	ω^6	ω^2	ω^4	0	ω	1	ω^5	ω^3

1.3 The miniquaternion system $\mathscr{Q}$

Next we construct a near-field $\mathscr{Q}$ of order 9 which is not left-distributive; $\mathscr{Q}$ will be called 'the miniquaternion system'.

The elements of $\mathscr{Q}$ are the same as those of $\mathscr{F}$; that is, the nine ordered pairs (a, b) with $a, b \in \mathscr{D}$. Furthermore, addition in $\mathscr{Q}$ is the same as in $\mathscr{F}$ (cf. Theorem 1.1.4):

$$(a_1, b_1)+(a_2, b_2) = (a_1+a_2, b_1+b_2).$$

That is, if we express the elements of $\mathscr{Q}$ as powers of ω the addition table is again Table 1.2.1.

While multiplication is of course different, we can define it in terms of multiplication in $\mathscr{F}$. We divide the non-zero elements

—that is, the set of eight powers $\omega^0, \omega^1, \omega^2, \ldots, \omega^7$ of ω—into two subsets $\mathscr{E}$ and $\mathscr{O}$:

$$\mathscr{E} = \{\omega^0, \omega^2, \omega^4, \omega^6\} = \{\text{even powers of } \omega\},$$

$$\mathscr{O} = \{\omega^1, \omega^3, \omega^5, \omega^7\} = \{\text{odd powers of } \omega\}.$$

In the field $\mathscr{F}$ multiplication is given by the rules:

$$0.\xi = \xi.0 = 0 \quad \text{for all} \quad \xi, \quad \text{and} \quad \omega^r.\omega^s = \omega^{r+s}$$

(indices taken modulo 8). The new multiplication $\otimes$ is defined as follows:

$$0 \otimes \xi = 0, \quad \xi \otimes 0 = 0 \qquad \text{for all } \xi$$

$$\omega^r \otimes \omega^s = \omega^r.\omega^s = \omega^{r+s} \qquad \text{if} \quad \omega^s \in \mathscr{E}$$

$$\omega^r \otimes \omega^s = \omega^{r*}.\omega^s = \omega^{3r}.\omega^s = \omega^{3r+s} \quad \text{if} \quad \omega^s \in \mathscr{O}.$$

Thus, for example,

$$\omega^2 \otimes \omega = \omega^6.\omega = \omega^7, \quad \text{but} \quad \omega \otimes \omega^2 = \omega.\omega^2 = \omega^3.$$

So $\otimes$ is not commutative. Also $\otimes$ is not left-distributive over $+$: from Table 1.2.1

$$\omega \otimes (\omega + 1) = \omega \otimes \omega^7 = \omega^3.\omega^7$$
$$= \omega^2,$$

whereas
$$\omega \otimes \omega + \omega \otimes 1 = \omega^3.\omega + \omega.\omega^0 = \omega^4 + \omega$$
$$= \omega^6.$$

The *miniquaternion system* $\mathscr{Q}$ consists of the nine elements (a, b), and the operations $+$, $\otimes$.

EXERCISE 1.3.1 Express $(1, 1)$ and $(1, -1)$ as powers of ω and hence determine the product $(1, 1) \otimes (1, -1)$.

THEOREM 1.3.1 *$\mathscr{Q}$ is a near-field of order* 9.

$\mathscr{Q}$, by definition, has nine elements. The operation $+$, being the same as in $\mathscr{F}$, is a commutative group operation. The operation $\otimes$ has identity

$$1 = \omega^0 \quad (\omega^r \otimes \omega^0 = \omega^r.\omega^0 = \omega^r; \quad \omega^0 \otimes \omega^r = \omega^0.\omega^r = \omega^r).$$

If $\omega^r \in \mathscr{E}$ then it has inverse ω^{-r} with respect to $\otimes$, and if $\omega^r \in \mathscr{O}$ then it has inverse ω^{-3r}. To show this, we remark first that, in $\mathscr{F}$, $\omega^{-r} = \omega^{8-r}$, so that

$$\omega^r \in \mathscr{E} \Rightarrow \omega^{-r} \in \mathscr{E} \quad \text{and} \quad \omega^{3r} \in \mathscr{E},$$

and

$$\omega^r \in \mathscr{O} \Rightarrow \omega^{-r} \in \mathscr{O} \quad \text{and} \quad \omega^{3r} \in \mathscr{O}.$$

Thus, if $\omega^r \in \mathscr{E}$, then

$$1 = \omega^{-r} . \omega^r = \omega^{-r} \otimes \omega^r,$$

$$1 = \omega^r . \omega^{-r} = \omega^r \otimes \omega^{-r},$$

while, if $\omega^r \in \mathscr{O}$, then

$$1 = \omega^{-r} . \omega^r = \omega^{-3r} \otimes \omega^r,$$

$$1 = \omega^{3r} . \omega^{-3r} = \omega^r \otimes \omega^{-3r}.$$

It remains to be proved that $\otimes$ is associative and right-distributive over $+$. If $\omega^t \in \mathscr{E}$,

$$(\omega^r + \omega^s) \otimes \omega^t = (\omega^r + \omega^s) . \omega^t = \omega^r . \omega^t + \omega^s . \omega^t$$

$$= \omega^r \otimes \omega^t + \omega^s \otimes \omega^t;$$

whereas, if $\omega^t \in \mathscr{O}$,

$$(\omega^r + \omega^s) \otimes \omega^t = (\omega^r + \omega^s)^3 . \omega^t = (\omega^{3r} + \omega^{3s}) . \omega^t$$

$$= \omega^{3r} . \omega^t + \omega^{3s} . \omega^t$$

$$= \omega^r \otimes \omega^t + \omega^s \otimes \omega^t.$$

If $\omega^s, \omega^t \in \mathscr{E}$,

$$(\omega^r \otimes \omega^s) \otimes \omega^t = \omega^{r+s} \otimes \omega^t = \omega^{r+s+t}$$

$$= \omega^r \otimes \omega^{s+t} \quad \text{as} \quad \omega^{s+t} \in \mathscr{E}$$

$$= \omega^r \otimes (\omega^s \otimes \omega^t).$$

If $\omega^s \in \mathscr{E}$ and $\omega^t \in \mathscr{O}$,

$$(\omega^r \otimes \omega^s) \otimes \omega^t = \omega^{r+s} \otimes \omega^t = \omega^{3r+3s+t}$$

$$= \omega^r \otimes \omega^{3s+t} \quad \text{as} \quad \omega^{3s+t} \in \mathscr{O}$$

$$= \omega^r \otimes (\omega^s \otimes \omega^t).$$

We leave to the reader the proofs of associativity when

$$\omega^s \in \mathscr{O}, \quad \omega^t \in \mathscr{E} \quad \text{and when} \quad \omega^s, \omega^t \in \mathscr{O}.$$

EXERCISE 1.3.2 Write the elements of $\mathscr{E}$ in the form $a+b\epsilon$.

EXERCISE 1.3.3 Show that

$$(\xi \otimes \eta)^* = \xi^* \otimes \eta^* \quad \text{for all} \quad \xi, \eta.$$

EXERCISE 1.3.4 Let η^i denote the inverse of η under $\otimes$ (and η^{-1} the inverse of η under ., as usual). Prove that

$$\xi \otimes \eta^i = \begin{cases} \xi . \eta^{-1} & \text{when} \quad \eta \in \mathscr{E}, \\ (\xi . \eta^{-1})^* & \text{when} \quad \eta \in \mathscr{O}. \end{cases}$$

EXERCISE 1.3.5 Consider $\xi = a + a'\epsilon$ as a row-vector

$$[a, a'] \quad (a, a' \in \mathscr{D}),$$

and let $$j_\xi = a^2 + a'^2 \quad (= \xi . \xi^*),$$

so that, unless $\xi = 0$, $j_\xi = \pm 1$. Also let

$$\mathbf{M}_\xi = \begin{bmatrix} a & a' \\ -a' & a \end{bmatrix},$$

$$\mathbf{J}_+ = \begin{bmatrix} 1 & 0 \\ 0 & 1 \end{bmatrix}, \quad \mathbf{J}_- = \begin{bmatrix} 1 & 0 \\ 0 & -1 \end{bmatrix}, \quad \mathbf{J}_\xi = \begin{bmatrix} 1 & 0 \\ 0 & j_\xi \end{bmatrix}.$$

Show that

(i) $$j_\xi = \begin{cases} 1 & \text{if} \quad \xi \in \mathscr{E}, \\ -1 & \text{if} \quad \xi \in \mathscr{O}, \end{cases}$$

(ii) $\xi . \eta = \xi \mathbf{M}_\eta = \xi \mathbf{J}_+ \mathbf{M}_\eta, \quad \xi^* . \eta = \xi \mathbf{J}_- \mathbf{M}_\eta,$

and $$\xi \otimes \eta = \xi \mathbf{J}_\eta \mathbf{M}_\eta \quad \text{for all} \quad \xi, \eta.$$

We are to derive a description of $\mathscr{Q}$ which does not explicitly involve $\mathscr{F}$ and which will clarify the name 'miniquaternion system'.

Recall that, in $\mathscr{F}$, $\epsilon^2 = -1$. That is -1 has a square root in $\mathscr{F}$. In a commutative field of odd order, a given non-zero element has either no square roots or exactly two square roots. (In $\mathscr{F}$, -1 has the square roots $\pm\epsilon$.) The situation is quite different in $\mathscr{Q}$. For

$$1 \otimes 1 = 1, \quad \omega \otimes \omega = \omega^2 \otimes \omega^2 = \omega^3 \otimes \omega^3 = \omega^4 (= -1),$$

$$\omega^4 \otimes \omega^4 = 1, \quad \text{and} \quad \omega^5 \otimes \omega^5 = \omega^6 \otimes \omega^6 = \omega^7 \otimes \omega^7 = \omega^4,$$

so that -1 has six square roots in $\mathscr{Q}$, namely $\omega, \omega^2, \omega^3, \omega^5, \omega^6, \omega^7$. Note that (since $\omega^4 = -1$) $\omega^5 = -\omega$, $\omega^6 = -\omega^2, \omega^7 = -\omega^3$. The resemblance to Hamilton's quaternion system is striking: there the elements $\pm i$, $\pm j$, $\pm k$ all have square -1. In fact the analogy goes further:

$$\omega \otimes \omega^2 = \omega^3 \quad \text{and} \quad \omega^2 \otimes \omega = \omega^7 = -\omega^3,$$

so that
$$\omega \otimes \omega^2 = -\omega^2 \otimes \omega = \omega^3;$$

and similarly
$$\omega \otimes \omega^3 = -\omega^3 \otimes \omega = -\omega^2,$$

$$\omega^2 \otimes \omega^3 = -\omega^3 \otimes \omega^2 = \omega.$$

The position is made clearer if we write say

$$\alpha = \omega, \quad \beta = \omega^6, \quad \gamma = \omega^7.$$

Then
$$\alpha \otimes \beta = \gamma = -\beta \otimes \alpha,$$

$$\beta \otimes \gamma = \alpha = -\gamma \otimes \beta,$$

$$\gamma \otimes \alpha = \beta = -\alpha \otimes \gamma,$$

and
$$\alpha \otimes \alpha = \beta \otimes \beta = \gamma \otimes \gamma = -1.$$

In fact, the multiplicative group of $\mathscr{Q}$ is isomorphic to the group with elements ± 1, $\pm i$, $\pm j$, $\pm k$ in the standard quaternion system.

From Table 1.2.1 we have $\omega^6 = \omega - 1$ and $\omega^7 = \omega + 1$, so that $\beta = \alpha - 1$ and $\gamma = \alpha + 1$. The nine elements of $\mathscr{Q}$ are thus

$$0,\ 1,\ -1,\ \alpha,\ -\alpha,\ \beta = \alpha - 1,$$
$$-\beta = -\alpha + 1,\ \gamma = \alpha + 1,\ -\gamma = -\alpha - 1.$$

The miniquaternion system $\mathscr{Q}$ is completely determined when it is described as a near-field with nine elements 0, ± 1, $\pm \alpha$, $\pm \beta$, $\pm \gamma$ subject to the relations:

$$\beta = \alpha - 1,\ \gamma = \alpha + 1,\ \alpha \otimes \alpha = \beta \otimes \beta = \gamma \otimes \gamma = -1,$$

$$a \otimes \beta = \gamma = -\beta \otimes \alpha,\ \beta \otimes \gamma = \alpha = -\gamma \otimes \beta,\ \gamma \otimes \alpha = \beta = -\alpha \otimes \gamma.$$

Some of these relations may be deduced from others using the near-field properties. For example,

$$\begin{aligned} \gamma\otimes\alpha &= (\alpha+1)\otimes\alpha && (\gamma = \alpha+1) \\ &= \alpha\otimes\alpha+1\otimes\alpha && \text{(right-distributive)} \\ &= \alpha\otimes\alpha+\alpha && \\ &= -1+\alpha && (\alpha\otimes\alpha = -1) \\ &= \beta && (\beta = \alpha-1). \end{aligned}$$

EXERCISE 1.3.6 Using the properties of finite near-fields, derive the relation $\alpha\otimes\beta = \gamma$ from the relations

$$\alpha\otimes\alpha = -1, \quad \beta = \alpha-1, \quad \gamma = \alpha+1.$$

We shall usually denote multiplication in $\mathscr{Q}$ by simple juxtaposition (as in $\mathscr{D}$ and $\mathscr{F}$).

For future reference we summarize some of the properties of $\mathscr{Q}$ (using $\xi\eta$ instead of $\xi\otimes\eta$, ξ^2 instead of $\xi\otimes\xi$, etc.):

THEOREM 1.3.2 *In the miniquaternion near-field*

$$\mathscr{Q} = \{0, \pm 1, \pm\alpha, \pm\beta, \pm\gamma\},$$

let

$$\mathscr{Q}^* = \{\pm\alpha, \pm\beta, \pm\gamma\} = \mathscr{Q}-\mathscr{D}.$$

Then:

(i) *Relations under addition*

$$\alpha-\beta = \beta-\gamma = \gamma-\alpha = 1, \quad \alpha+\beta+\gamma = 0.$$

(ii) *Relations under multiplication*

$$\alpha^2 = \beta^2 = \gamma^2 = -1, \quad \textit{that is} \quad \xi\in\mathscr{Q}^* \Rightarrow \xi^2 = -1,$$

$$\alpha\beta\gamma = -1,$$

$$\beta\gamma = -\gamma\beta = \alpha, \quad \gamma\alpha = -\alpha\gamma = \beta, \quad \alpha\beta = -\beta\alpha = \gamma,$$

that is

$$\xi,\eta\in\mathscr{Q}^* \textit{ and } \xi \neq \pm\eta \Rightarrow \xi\eta = -\eta\xi.$$

Since $\xi\eta = \eta\xi$ if either $\xi\in\mathscr{D}$ or $\eta\in\mathscr{D}$, any two elements of $\mathscr{Q}$ either commute or 'anti-commute' under multiplication.

1.4 The automorphism group of $\mathscr{Q}$

Let $\sigma \in \mathscr{Q}^* = \{\pm\alpha, \pm\beta, \pm\gamma\}$. Then each element of $\mathscr{Q}$ may be written uniquely in the form $a+b\sigma$ with $a, b \in \mathscr{D}$ (Theorem 1.1.3). There are six elements σ in $\mathscr{Q}^*$, and each of these determines a permutation

$$a+b\alpha \to a+b\sigma$$

of the set $\mathscr{Q}$. For example, if $\sigma = -\alpha$,

$$(0, 1, -1, \alpha, -\alpha, \beta, -\beta, \gamma, -\gamma) \to (0, 1, -1, -\alpha, \alpha, -\gamma, \gamma, -\beta, \beta).$$

If we write the permutation corresponding to σ as $\mathscr{S}: \xi \to \mathscr{S}(\xi)$ we can establish the properties

$$\text{(i)} \quad \mathscr{S}(\xi+\eta) = \mathscr{S}(\xi)+\mathscr{S}(\eta),$$

$$\text{(ii)} \quad \mathscr{S}(\xi\eta) = \mathscr{S}(\xi)\mathscr{S}(\eta) \quad \text{for all} \quad \xi, \eta.$$

Such a permutation is called an automorphism of $\mathscr{Q}$ (isomorphism of $\mathscr{Q}$ onto itself). The importance of an automorphism $\mathscr{T}$ is that if we establish algebraic relations between certain elements expressed in terms of say α, then the same relations hold when α is replaced by $\mathscr{T}(\alpha)$.

The proof of (i) for $\mathscr{S}$ is straightforward. To prove (ii) we observe first that since the product of any two members of $\mathscr{Q}$ is another member of $\mathscr{Q}$, if we are given $a, a', b, b' \in \mathscr{D}$, there are unique elements c, d of $\mathscr{D}$ such that

$$(a+b\alpha)(a'+b'\alpha) = c+d\alpha.$$

In calculating c, d the only properties of α needed are that α commutes with each element of $\mathscr{D}$ and $\alpha^2 = -1$. Consequently if σ is any of the six members of $\mathscr{Q}^*$ we have also

$$(a+b\sigma)(a'+b'\sigma) = c+d\sigma.$$

Now, let $\mathscr{S}(x+y\alpha) = x+y\sigma$, then

$$\mathscr{S}\{(a+b\alpha)(a'+b'\alpha)\} = \mathscr{S}(c+d\alpha) = c+d\sigma$$

and $\quad \mathscr{S}(a+b\alpha)\mathscr{S}(a'+b'\alpha) = (a+b\sigma)(a'+b'\sigma) = c+d\sigma.$

Thus $\mathscr{S}$ has property (ii).

EXERCISE 1.4.1 Show that:

(i) the automorphisms of $\mathscr{Q}$ form a group under composition,

(ii) every automorphism of $\mathscr{Q}$ leaves $0, 1, -1$ fixed,

(iii) apart from the identity, no automorphism of $\mathscr{Q}$ leaves α fixed.

NOTATION 1.4.1 Aut $(\mathscr{Q})$: the automorphism group of $\mathscr{Q}$.

We examine the structure of Aut $(\mathscr{Q})$. Let $\mathscr{A}, \mathscr{B}, \mathscr{C}$ be respectively the automorphisms $\mathscr{S}$ which map α to $-\alpha$, $-\gamma$, $-\beta$. It is easily checked that $\mathscr{B}(\beta) = -\beta, \mathscr{C}(\gamma) = -\gamma$. Now an automorphism is uniquely determined by its effect on α, and

$$\mathscr{A}\mathscr{C}\mathscr{A}(\alpha) = \mathscr{A}\mathscr{C}(-\alpha) = \mathscr{A}(\beta) = -\gamma = \mathscr{B}(\alpha),$$

so $\mathscr{A}\mathscr{C}\mathscr{A} = \mathscr{B}$. Similarly $\mathscr{C}\mathscr{A}\mathscr{C} = \mathscr{B}$. The automorphisms

$$1, \mathscr{A}, \mathscr{C}, \mathscr{A}\mathscr{C}, \mathscr{C}\mathscr{A}, \mathscr{B} = \mathscr{A}\mathscr{C}\mathscr{A} = \mathscr{C}\mathscr{A}\mathscr{C},$$

since they map α respectively to α, $-\alpha$, $-\beta$, γ, β, $-\gamma$, are distinct. They are, therefore, the six automorphisms $\mathscr{S}$. (If there were two automorphisms $\mathscr{U}$ and $\mathscr{V}$ mapping α to the same element σ then $\mathscr{U}^{-1}\mathscr{V}$ would fix α, implying $\mathscr{U}^{-1}\mathscr{V} = 1$, $\mathscr{U} = \mathscr{V}$.) We have proved:

THEOREM 1.4.1 *$\mathscr{Q}$ has exactly six automorphisms, one mapping α to any given $\sigma \in \mathscr{Q}^*$.* Aut $(\mathscr{Q})$ *is generated by $\mathscr{A}$ and $\mathscr{C}$, where* $\mathscr{A}(\alpha) = -\alpha, \mathscr{C}(\gamma) = -\gamma$.

In fact, Aut $(\mathscr{Q})$ is isomorphic to S_3, the group of permutations on three objects.

EXERCISE 1.4.2 Show that Aut $(\mathscr{F})$ contains only the identity permutation and the conjugation permutation $\xi \to \xi^*$ (Definition 1.2.1).

EXERCISE 1.4.3. Verify that, when $\mathscr{Q}$ is defined in terms of $\mathscr{F}$ (as on page 9), $\xi \to \xi^*$ is the automorphism $\mathscr{B}$ of $\mathscr{Q}$ such that $\mathscr{B}(\beta) = -\beta$.

The last result may be expressed in the form:

$$\alpha^* = -\gamma, \quad \beta^* = -\beta, \quad \gamma^* = -\alpha.$$

The anomalous role played by β is a consequence of the definitions $\beta+1 = \alpha = \omega = 1-\epsilon$, which imply that $\beta = -\epsilon$ and

$$\beta^* = (-\epsilon)^* = \epsilon = -\beta,$$

whereas $\alpha^* = 1+\epsilon = -\gamma$, $\gamma^* = -1+\epsilon = -\alpha$.

1.5 The solution of equations in $\mathcal{Q}$

We have already seen that the equation $x^2 = -1$ has six roots in $\mathcal{Q}$. Equations over $\mathcal{Q}$ therefore behave rather differently from equations over a commutative field. Indeed, since multiplication in $\mathcal{Q}$ is neither commutative nor left-distributive over addition, complications may arise in solving (for x) even a 'linear' equation of the form

$$\rho x \rho' + \sigma x \sigma' = \tau \text{ (all coefficients non-zero).}$$

Using the right-distributive law, we obtain the equivalent form $[\rho x \rho'(\sigma')^{-1} + \sigma x]\sigma' = \tau$, or $\rho x \rho'' + \sigma x = \tau''$, where

$$\rho'' = \rho'(\sigma')^{-1}, \quad \tau'' = \tau(\sigma')^{-1}.$$

By the change of variable $y = \rho x$ this becomes $y\rho'' + \sigma\rho^{-1}y = \tau''$, or, writing $\sigma'' = \sigma\rho^{-1}$,

$$y\rho'' + \sigma'' y = \tau''.$$

We wish therefore to solve equations of the type

$$x\rho + \sigma x = \tau.$$

If $\rho \in \mathcal{D}$ then, replacing $x\rho$ by ρx, we may use the right-distributive law to obtain, as the solution of $\rho x + \sigma x = \tau$,

$$x = (\rho+\sigma)^{-1}\tau$$

(unless of course $\rho + \sigma = 0$, when there is no solution unless $\tau = 0$).

If $\sigma = 1$ in $x\rho + \sigma x = \tau$ then we are concerned with the equation $x\rho + x = \tau$. For a solution to exist we require $\rho \neq -1$. The case $\rho = 1$ is trivial (the solution is $x = -\tau$), so we may assume that $\rho \in \mathcal{Q}^*$. Now we use the automorphisms $\mathcal{S}$ of $\mathcal{Q}$. $x\rho + x = \tau$ if and only if $\mathcal{S}(x)\mathcal{S}(\rho) + \mathcal{S}(x) = \mathcal{S}(\tau)$, so that, replacing x by

$y = \mathscr{S}(x)$, and choosing $\mathscr{S}$ so that $\mathscr{S}(\rho) = \alpha$, we are required to to solve

$$y\alpha + y = \tau' \quad (\tau' = \mathscr{S}(\tau)).$$

Now when

$$\begin{array}{rccccc} y = 0 & \pm 1 & \pm\alpha & \pm\beta & \pm\gamma, \\ y\alpha + y = 0 & \pm\gamma & \pm\beta & \pm 1 & \mp\alpha. \end{array}$$

Thus there is exactly one solution for the equation $y\alpha + y = \tau'$. We have proved

THEOREM 1.5.1 *For given values of ρ, τ with $\rho \neq -1$, the solution in $\mathscr{Q}$ of $x\rho + x = \tau$ exists and is unique.*

EXERCISE 1.5.1 (i) Using a suitable automorphism, write down the values of $-y\beta + y$ from the above table for $y\alpha + y$.

(ii) Show that the equation $x\rho + x\sigma = \tau$ has a unique solution x whenever $\sigma \neq -\rho$.

Returning to the general equation

$$x\rho + \sigma x = \tau,$$

we may multiply this on the right by τ^{-1} to obtain (using right-distributivity)

$$x\rho\tau^{-1} + \sigma x\tau^{-1} = 1,$$

or, putting $y = x\rho\tau^{-1}$,

$$y + \sigma y\tau\rho^{-1}\tau^{-1} = 1,$$

which is an equation of the form

$$y + \theta y\phi = 1.$$

If $\theta = \pm 1$ or $\phi = \pm 1$ we obtain an equation solvable by the methods established above; we assume therefore that $\theta, \phi \in \mathscr{Q}^*$. Consider first the expression $z + \alpha z\beta$. We have the table

$$\begin{array}{rccccc} z = 0 & \pm 1 & \pm\alpha & \pm\beta & \pm\gamma, \\ z + \alpha z\beta = 0 & \pm\beta & \pm 1 & \mp 1 & \pm\beta. \end{array}$$

Thus

$z + \alpha z\beta = 0$ has one solution, $z = 0$,

$z + \alpha z\beta = \pm 1$ have two solutions each, $z = \pm\{\alpha, -\beta\}$,

$z + \alpha z\beta = \pm\beta$ have two solutions each, $z = \pm\{1, \gamma\}$,

$z + \alpha z\beta = \pm\alpha$ or $\pm\gamma$ have no solutions.

EXERCISE 1.5.2 (i) Verify that, for $z \neq 0$,

$$z - \alpha z \beta = \pm \alpha \quad \text{or} \quad \pm \gamma$$

and $$z \pm \alpha z \alpha = 0 \text{ (4 times)}, \quad -z \text{ (4 times)}.$$

(ii) Given a form $y + \theta y \phi$ ($\theta, \phi, \in \mathcal{Q}^*$), show that there exists an automorphism, $z = \mathcal{S}(y)$, which transforms it to one of

$$z + \alpha z \beta, \quad z - \alpha z \beta, \quad z + \alpha z \alpha, \quad z - \alpha z \alpha.$$

As a result of the above we have

THEOREM 1.5.2 *The equation $x + \theta x \phi = 1$, where $\theta, \phi \in \mathcal{Q}^*$, has*

2 solutions when $\theta - \phi = \pm 1$,

no solution when $\theta + \phi = \pm 1$,

no solution when $\theta = \phi$,

1 solution ($x = -1$) when $\theta = -\phi$.

EXERCISE 1.5.3 (i) Using Theorems 1.5.1, 2, discuss the solutions of $x\rho + \sigma x = \tau$ for various values of ρ, σ, τ.

(ii) Reduce $\rho x + \sigma(\sigma' x + \kappa) = \tau$ to the form $\lambda y + \mu(y+1) = \nu$. Discuss the solutions of $\alpha y + \beta(y+1) = \nu$ for various values of ν.

(iii) Show that $x\alpha + x\beta + x = \kappa$ has three solutions when $\kappa = 0$ or α or $-\alpha$, but no solution for any other value of κ.

As we have shown, each of the equations $\rho x + \sigma x = \tau$ and $x\rho + x\sigma = \tau$ has a unique solution, provided $\sigma \neq -\rho$, whereas the equation $x\rho + \sigma x = \tau$ may have several solutions, one solution or no solution.

The situation is still more complicated in the case of pairs of 'linear' equations in two variables x and y. But if the equations are to correspond to lines in a geometrical system, there must be at most one solution (except when the two equations represent the same line).

Guided by forms usually adopted for the equations of lines in the Cartesian analysis of a Euclidean plane, we select first two forms of equations which provide unique solutions, and follow this with an investigation of some other forms.

THEOREM 1.5.3 *Each of the following pairs of equations in x and y has, provided $\mu \neq \mu'$, a unique solution in $\mathcal{Q}$:*

(i) $y = \mu x + \nu,$

$\quad y = \mu' x + \nu',$

(ii) $y = x\mu + \nu,$

$\quad y = x\mu' + \nu'.$

The two equations imply

(i) $\mu x - \mu' x = \nu' - \nu \Rightarrow x = (\mu - \mu')^{-1}(\nu' - \nu)$,

(ii) $x\mu - x\mu' = \nu' - \nu$. This equation has a unique solution (Exercise 1.5.1).

EXERCISE 1.5.4 Show that the pair of equations

$$x\lambda + y\mu = \nu,$$

$$x\rho + y\sigma = \tau,$$

has a unique solution when $\mu, \sigma \neq 0$ and $\lambda\mu^{-1} \neq \rho\sigma^{-1}$.

Consider now the equations

$$\lambda x + \mu y = \nu,$$

$$\rho x + \sigma y = \tau,$$

with all coefficients non-zero. This pair is analogous to the pair of Exercise 1.5.4. We have $\mu y = \nu - \lambda x$, $\sigma y = \tau - \rho x$ so that the equations can be solved uniquely if

$$\sigma\mu^{-1}(\nu - \lambda x) = \tau - \rho x$$

has a unique solution. Multiplying on the right by ν^{-1} we obtain

$$\rho x \nu^{-1} + \sigma\mu^{-1}(1 - \lambda x \nu^{-1}) = \tau\nu^{-1};$$

substituting u for $-\lambda x \nu^{-1}$ we have an equation of the form

$$\theta u + \phi(1+u) = \psi.$$

Take in particular $\theta = \alpha$, $\phi = \beta$:

$$\begin{array}{rcccccccccc} u & = & 0 & 1 & -1 & \alpha & -\alpha & \beta & -\beta & \gamma & -\gamma, \\ \alpha u + \beta(1+u) & = & \beta & 1 & -\alpha & \beta & -1 & 0 & \beta & -\alpha & -\alpha. \end{array}$$

The equation $\alpha u + \beta(1+u) = \psi$ has therefore one solution when $\psi = 0$, 1 or -1, three solutions when $\psi = -\alpha$ or β and otherwise no solutions.

EXERCISE 1.5.5 Find a pair of equations

$$y = x\mu + \nu,$$

$$y = \rho x + \sigma,$$

having precisely two solutions.

PART II

FIELD-PLANES

We describe some general properties of projective planes, and show how an arbitrary field determines a projective plane. This plane we call a *field-plane*, or, if the field is finite, a *Galois plane*. The case of a commutative field, particularly the Galois field of order 9, is considered in some detail, because the Galois plane of order 9, besides being of interest in itself, provides clues to the type of properties to be sought in the miniquaternion planes.

Chapter 2

PROJECTIVE PLANES

2.1 Elementary properties, subplanes

Suppose we adjoin to the real Euclidean plane a 'line at infinity', which consists of one 'point at infinity' ('ideal point') corresponding to each direction, opposite directions being counted as the same. The new system has the advantage that not only is there a line joining any two points, there is also a point common to any two lines.

If we concentrate on incidence rather than metric properties, the following generalization arises naturally:

Definition 2.1.1 A *projective plane* is a system of points and lines related by incidence (point P lies on line l, or line l passes through point P) satisfying the axioms:

1. Given any two distinct points, there is exactly one line which passes through both.
2. Given any two distinct lines, there is exactly one point which lies on both.
3. There exists a set of four points no three of which lie on a single line.

The purpose of the third axiom is to exclude the following systems of points and lines which are of little interest: (i) a set $\{A_i\}$ of collinear points, a point V not collinear with them, the line containing $\{A_i\}$ and the set of lines $\{VA_i\}$, (ii) a set of collinear points and a set of lines concurrent in one of them, (iii) any subset of systems (i) or (ii) which satisfies axioms 1 and 2 either meaningfully or vacuously.

Definition 2.1.2 (i) $AB\,(=BA)$ or $\langle \mathbf{a}, \mathbf{b} \rangle$: the unique line containing the pair of distinct points A, B or $\mathbf{a}$, $\mathbf{b}$. AB is the *join* of A and B.

(ii) $l \cap m$: the unique point of intersection of the two distinct lines l, m.

(iii) *Regular quadrangle*: a set of four points of which no three are collinear. If $ABCD$ is a regular quadrangle, the *sides* of the quadrangle are the six lines AB, AC, AD, BC, BD, CD; the *diagonal triangle* of the quadrangle has *vertices*

$$BC \cap AD, \quad CA \cap BD, \quad AB \cap CD$$

and *sides* which are the joins of pairs of these.

(iv) *Regular quadrilateral*: a set of four lines of which no three are concurrent.

Little can be proved from these axioms alone. Usually some further axiom is added, for instance the Desargues proposition (of which a detailed account is given in, for example, Room, 1967, p. 24) or a weak form of this proposition. We need, however, no further axiom to prove the following two results.

THEOREM 2.1.1 *From a projective plane* Π *a new system* Π^D *is constructed thus: the points of* Π^D *are the lines of* Π *and the lines of* Π^D *are the points of* Π, *a point and line being incident in* Π^D *if and only if they are incident (as line and point respectively) in* Π. *Then* Π^D *is a projective plane.*

Π^D satisfies axiom 1 since Π satisfies axiom 2, and axiom 2 since Π satisfies axiom 1. If $ABCD$ is a regular quadrangle in Π, then no three of the lines AB, BC, CD, DA are concurrent, and therefore they form a regular quadrangle in Π^D. Thus Π^D satisfies axiom 3.

DEFINITION 2.1.3 Π^D, as described in Theorem 2.1.1, is called the *dual plane* of Π.

Two of the four 91-point planes which are the subject of this book are their own duals and the other two are duals of each other.

THEOREM 2.1.2 (i) *There is a* 1–1 *correspondence between the set of points on any line* l *and the set of points on any other line* m.

(ii) *There is a* 1–1 *correspondence between the set of points on* l *and the set of lines through any point* P.

Suppose $P \notin l$ (P does not lie on l). Then, since each line through P meets l in one and only one point, and each point L of l determines a single line PL through P, there is a 1–1 correspondence between the points L of l and the lines PL through P. Let l, m be two distinct lines. Choose a point Q with $Q \notin l$, $Q \notin m$. There is a 1–1 correspondence between the points of l and lines through Q; and another between points of m and lines through Q. This proves (i). To prove (ii) we need only consider now the case: $P \in l$. Choose a line n with $P \notin n$. There is a 1–1 correspondence between points on l and points on n; also between points on n and lines through P.

Next, suppose that in some projective plane Π there are only finitely many, say $k+1$, points on some line. By Theorem 2.1.2, there is the same number of points on each line of Π and $k+1$ lines through each point of Π. Choose a point P in Π. Every point $\neq P$ lines on exactly one of the $k+1$ lines through P, thus there are altogether $(k+1)k+1 = k^2+k+1$ points in Π. Similarly there are k^2+k+1 lines in Π.

DEFINITION 2.1.4 A projective plane such that one (and therefore each) of its lines contains a finite number $k+1$ of points is called a *plane of order* k.

By the above we have:

THEOREM 2.1.3 *In a finite projective plane of order k there are altogether k^2+k+1 points and k^2+k+1 lines.*

We are to prove immediately that there is a plane containing only seven points ($k = 2$). This plane was first constructed by G. Fano in 1891 and is now customarily known as the *Fano plane*.

Let $\{A, B, C, D, E, F, G\}$ be the set of seven elements (the points) which are to form, if possible, a plane of order 2. In order that they shall do so we have to be able to arrange them in seven sets of three (the lines) in such a way that there are three lines containing each point. Let the three lines through A be $\{A, B, D\}$, $\{A, C, G\}$, $\{A, E, F\}$. We require the line joining B and C to meet $\{A, E, F\}$ in either E or F. Relabel E and F if necessary so that

$BC \cap AE$ (the point of intersection of the lines BC and AE) is E. Then, of necessity,

$$BG \cap AE = F, \quad DC \cap AE = F, \quad DG \cap AE = E.$$

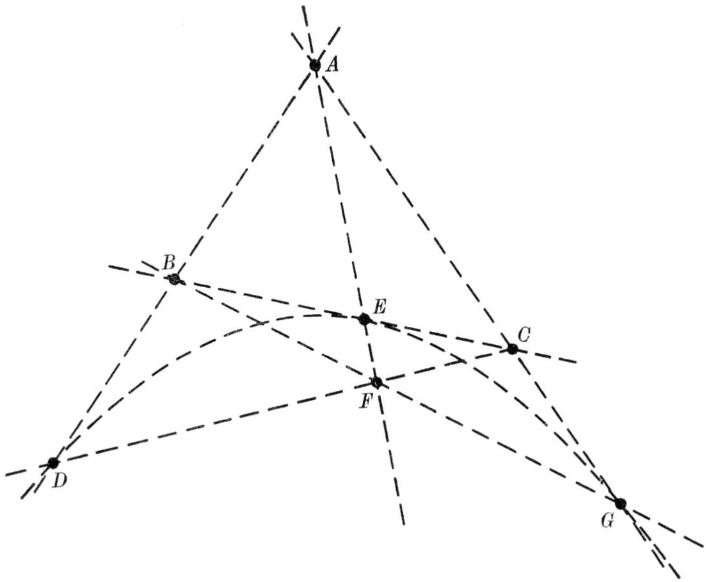

Diagram 2.1 (*a*)

Thus, if there is a plane of order 2, its points may be labelled so that the lines are given by the columns in the table

A	B	C	D	E	F	G
B	C	D	E	F	G	A
D	E	F	G	A	B	C

It is easily verified that this table does indeed describe a projective plane of order 2.

We have shown, in the course of constructing a plane of order 2, that all such planes may be represented by the same table, in other words that they are all 'essentially the same' or 'isomorphic'.

DEFINITION 2.1.5 Two projective planes are *isomorphic* if there is a 1–1 correspondence between the points of one and the points of the other, and between the lines of one and the lines of the other, such that

$$(A \leftrightarrow A' \quad \text{and} \quad l \leftrightarrow l') \Rightarrow (A \in l \Leftrightarrow A' \in l').$$

THEOREM 2.1.4 *There exists a plane of order* 2, *and all (projective) planes of order* 2 *are isomorphic.*

We often say, simply: there is a unique plane of order 2.

In the Fano plane the diagonal triangle of any regular quadrangle (for example, *ABCF*) collapses into a set of three collinear points (*DEG*), namely the remaining three points in the plane.

The pattern of incidences is clarified if we replace the letters $A, \ldots, G$ by the digits $0, \ldots, 6$. We shall use as a standard table of incidences in the Fano plane:

Table 2.1.1

0	1	2	3	4	5	6
1	2	3	4	5	6	0
3	4	5	6	0	1	2

The line containing the three points $r, r+1, r+3$ (mod 7) could be referred to as the line r.

The Fano plane occurs as a 'subplane' of many larger planes.

DEFINITION 2.1.6 A *subplane* of a plane Π is a system Π' of points and lines in Π which itself satisfies axioms 1, 2, 3 of Definition 2.1.1.

The problem of determining for a given plane (other than a field plane) the subplanes and their intersection properties is at present largely unsolved. The following theorem, due to R. H. Bruck (see M. Hall, 1959, p. 398), provides a restriction on the range of possible orders of subplanes in the finite case.

THEOREM 2.1.5 (BRUCK'S THEOREM) *If* Π' *is a subplane of a finite plane* Π *and the orders of* Π, Π' *are* n, m *respectively, then unless* $\Pi' = \Pi$ *either*

$$\text{(i) } m^2+m \leqslant n \quad \textit{or} \quad \text{(ii) } m^2 = n.$$

Suppose $\Pi' \neq \Pi$ and let l be a line of Π'. Select a point P on l which is not in Π'. There are $(m^2+m+1)-(m+1) = m^2$ points $Q_1, \ldots, Q_{m^2}$ in Π' which do not lie on l. If P, Q_i, Q_j $(i \neq j)$ were collinear then the line PQ_i would contain two points of Π' and therefore lie in Π', so that P $(= l \cap PQ_i)$ would lie in Π', which it does not. Thus $PQ_1, \ldots, PQ_{m^2}$ are m^2 distinct lines through P. As l also passes through P and there are only $n+1$ lines through P, $n+1 \geqslant m^2+1$; that is, $m^2 \leqslant n$. Suppose $m^2+1 < n+1$. Then there must exist a line which passes through P and contains no point of Π'. This line meets the m^2+m+1 lines of Π' in m^2+m+1 distinct points, since the intersection of two lines of Π' is a point of Π'. It follows that

$$n+1 \geqslant m^2+m+1,$$

that is $m^2+m \leqslant n$.

From this theorem we can deduce for instance that the only possible orders for a proper subplane of a plane of order 9 are 2 and 3; also that the plane of order 3 (see § 3.1) can have no Fano subplane.

In proving Bruck's Theorem we showed that, through the point P (chosen as above), there is a line which contains no points of Π', if and only if $m^2+m \leqslant n$. Since any line which contains no point of Π' necessarily meets any line of Π' in a point of Π which is not in Π', we deduce:

THEOREM 2.1.6 *If* Π' *is a subplane of order* m *in a finite plane* Π *of order* $n = m^2$, *then each line of* Π *contains at least one point of* Π'.

EXERCISE 2.1.1 Show that, under the hypotheses of Theorem 2.1.6, each point of Π lies on at least one line of Π'.

We conclude this section by describing the process of 'completing a regular quadrangle' which lies at the heart of many investigations of the structure of projective planes.

DEFINITION 2.1.7 *Completion of a quadrangle*: $ABCD$ is a regular quadrangle. The following succession of sets of elements is constructed: (i l) the line joining each of the six pairs of points, (i p) the point common to each of the three opposite pairs of these lines, (ii l) the lines joining the pairs of these points, (ii p) the six points in which these lines meet sides of the quadrangle, (iii l) the line joining each pair of points so far constructed and not already joined, (iii p) the point of intersection of every pair of lines so far constructed and of which the common point has not been named, (iv l) ....

On completing operation (ii p), if the quadrangle is not in a Fano plane, we shall have constructed the configuration of thirteen points and nine lines in Diagram 2.1 (b).

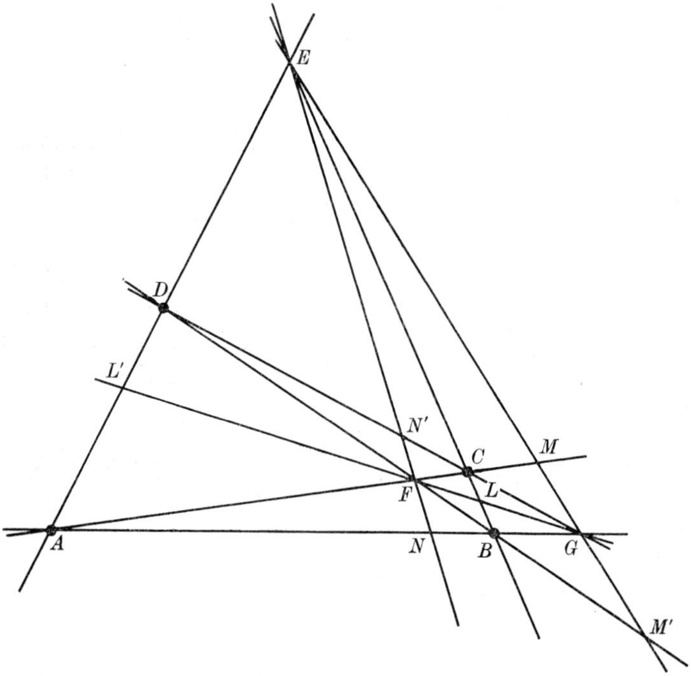

Diagram 2.1 (b)

If the plane is finite (that is, consists of finitely many points) then the process is finite and the resulting configuration is either the whole plane or some proper subplane. For example, the completion of any regular quadrangle in the Fano plane is the whole plane.

2.2 Collineations in a projective plane

Since points and lines are the basic objects of a projective plane, the point transformations of greatest interest are those which carry lines to lines.

DEFINITION 2.2.1 A *collineation* of a plane Π is a 1–1 transformation $\mathscr{T}$ of the set of points of Π onto itself, such that $\mathscr{T}$ maps lines onto lines.

EXERCISE 2.2.1 Show that the set of all collineations of a projective plane is a group under composition.

A type of collineation which involves a particularly simple construction is the following:

DEFINITION 2.2.2 *Central collineation*: a collineation $\mathscr{T}$ such that there is a point P, *the centre*, with the property that

$$\mathscr{T}(P) = P \quad \text{and, for every other point } X, \quad \mathscr{T}(X) \in PX.$$

A collineation $\mathscr{T}$ with centre P induces a permutation of the points on each line p which passes through P, since

$$(X \in p \quad \text{and} \quad p \supset P) \Rightarrow \mathscr{T}(X) \in p.$$

Thus although $\mathscr{T}$ does not in general fix (that is, map to itself) every point of such a line p, it fixes the line p. Hence a central collineation fixes every line through P; dually we can define:

DEFINITION 2.2.3 *Axial collineation*: a collineation such that there is a line, the *axis*, every point of which is fixed.

Later we are to prove that every central collineation is an axial collineation, and conversely. There are two different types of central collineations, those in which the centre and axis are 'disjoint' (that is, the centre does not lie on the axis), and those in which the centre lies on the axis.

Many of the transformations of elementary geometry are central collineations; displacements (translations), the reflection in a line l, dilatations. Diagrams 2.2(a), (b), (c) show the relation between the centre, axis, a pair of points X, Y and their images X', Y' for these collineations.

In each case it is easily seen that given the centre and axis, and given that X is mapped to X', the image Y' of Y is constructible once Y is chosen. For the translation Y' is obtained by completing a parallelogram. The segments XX' and YY' will

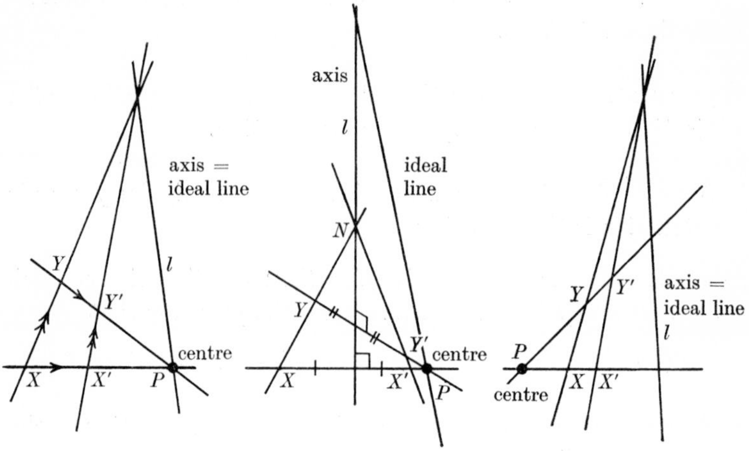

(a) Translation (b) Reflection in l (c) Dilatation

Diagram 2.2

have the same length and direction. For the reflection in l we require first that $XX' \perp l$ and that X, X' be the same distance from l, on opposite sides of l; given this then Y' is obtained by dropping the perpendicular from Y to l and taking the point along this which is the same distance from l as Y, but on the other side. Metrical concepts may be avoided here by using the following incidence construction (a construction which is valid for the translations and dilatations also). Call the centre P and the axis l, and let $N = XY \cap l$; then $Y' = X'N \cap PY$. For the line XN is mapped to the line $X'N$ since X is mapped to X' and N is fixed; also the line PY is fixed since P is the centre; and $Y = XN \cap PY$. This argument applies to any central collineation having an axis, consequently:

THEOREM 2.2.1 *A central collineation which has an axis is uniquely determined, if it exists, by the following elements: its centre P; its axis l; a point X such that $X \neq P, X \notin l$; and the image X' of X.*

The 'identity transformation' 1 (which fixes every point of the plane) has an axis. In fact, every line is an axis for it.

THEOREM 2.2.2 *Every non-trivial central collineation has one and only one axis.*

Let $\mathscr{T} \neq 1$ be a central collineation with centre P. If there is a fixed line not passing through P it is an axis, since every line through P is a fixed line.

Now $\mathscr{T}$ has at most one axis. For if $\mathscr{T}$ had two axes l and m, then every line not through $l \cap m$ would be a fixed line (joining a fixed point on l and a fixed point on m); but every point apart from $l \cap m$ is the intersection of two such lines, and so $\mathscr{T}$ would fix every point in the plane, contradicting $\mathscr{T} \neq 1$.

Thus every fixed line, except possibly one, passes through P. There must therefore be a line h which is not fixed. Let

$$A = h \cap \mathscr{T}(h); \quad \text{then} \quad A \neq P.$$

Since $\mathscr{T}(A) \in PA$, A is a fixed point. Let $B \in PA$, $B \neq P, A$. There exists a line k through B which is not a fixed line. Let $C = k \cap \mathscr{T}(k)$. C is a fixed point and $C \neq P, A$; also AC is a fixed line.

If $AC \not\supset P$ then AC is an axis. On the other hand if $AC \supset P$, then $B = C$ and the three collinear points P, A, B are fixed. If the plane is of order 2, then PAB is an axis. If not, let

$$B' \in AP, \quad B' \notin \{P, A, B\},$$

let k' be a line through B' which is not fixed, and let

$$C' = k' \cap \mathscr{T}(k').$$

If $C' \notin PA$, then $C'A$ and $C'B$ are distinct axes, which we have seen is not possible, and therefore $C' = B'$ and B' is a fixed point. It follows that every point of PA is fixed, and PA is an axis. Thus every non-trivial central collineation has one and only one axis.

EXERCISE 2.2.2 Prove that every axial collineation $\mathscr{T} \neq 1$ has exactly one centre.

As we have seen, the axis of a central collineation sometimes does, and sometimes does not, pass through the centre.

DEFINITION 2.2.4 *Homology*: a central collineation with non-incident centre and axis.

Elation: a central collineation with incident centre and axis.

(P, l)-collineation: a central collineation with an assigned centre P and an assigned axis l.

EXERCISE 2.2.3 Prove, for a given point P and a given line l, that the product of two (P, l)-collineations is a (P, l)-collineation, and that the set of (P, l)-collineations forms a group under composition.

DEFINITION 2.2.5 *(P, l)-transitive plane*: a plane is (P, l)-transitive for a given point P and a given line l if for each pair of distinct points $\{X, X'\}$, disjoint from $\{P, l\}$ but collinear with P,† there exists a (P, l)-collineation mapping X to X'.

(P, l)-transitivity plays an important role in the investigation of the structure of a projective plane, and the set of point-line pairs (P, l), for which a plane is (P, l)-transitive provides a basis for the classification of projective planes. In the following theorem we relate this set to the collineations of the plane.

THEOREM 2.2.3 *If for a given pair $\{P, l\}$ a plane Π is (P, l)-transitive, and $\mathscr{T}$ is a collineation of Π, then Π is also $(\mathscr{T}P, \mathscr{T}l)$-transitive.*

Let $P_1 = \mathscr{T}P$, $l_1 = \mathscr{T}l$; take any points X_1, X_1' such that P_1, X_1, X_1' are collinear and $\{X_1, X_1'\}$ and $\{P_1, l_1\}$ are disjoint, and let $X = \mathscr{T}^{-1}X_1$, $X' = \mathscr{T}^{-1}X_1'$. Since $\mathscr{T}^{-1}P_1 = P$, $\mathscr{T}^{-1}l_1 = l$ and $\mathscr{T}^{-1}$ is a collineation, the points P, X, X' are collinear and $\{X, X'\}$ is disjoint from $\{P, l\}$. So there is a unique (P, l)-collinea-

† That is: $X \neq P$, $X' \neq P$ and $X \notin l$, $X' \notin l$ and P, X, X' are collinear.

tion, $\mathscr{A}$ say, such that $\mathscr{A}X = X'$, and therefore there is a collineation $\mathscr{A}_1 = \mathscr{T}\mathscr{A}\mathscr{T}^{-1}$ such that

$$\mathscr{A}_1 X_1 = \mathscr{T}\mathscr{A}\mathscr{T}^{-1}X_1 = \mathscr{T}\mathscr{A}X = \mathscr{T}X' = X_1'.$$

Let Z_1 be any point on l_1. Then $\mathscr{T}^{-1}Z_1$ lies on l, the axis of $\mathscr{A}$, and so

$$\mathscr{A}_1 Z_1 = \mathscr{T}\mathscr{A}\mathscr{T}^{-1}Z_1 = \mathscr{T}\mathscr{T}^{-1}Z_1 = Z_1.$$

Thus, l_1 is an axis for $\mathscr{A}_1$. Similarly, P_1 is a centre for $\mathscr{A}_1$.

A criterion for (P,l)-transitivity when Π is finite is given by:

THEOREM 2.2.4 *Suppose* Π *is finite, of order* n.

(i) *If* $P \in l$ *and there are* n *distinct* (P,l)*-elations, then* Π *is* (P,l)*-transitive.*

(ii) *If* $P \notin l$ *and there are* $n-1$ *distinct* (P,l)*-homologies, then* Π *is* (P,l)*-transitive.*

To prove (i), choose a line $h\ (\neq l)$ through P, and a point H on $h\ (H \neq P)$. The n elations map H to the n points of h distinct from P (Theorem 2.2.1). Thus there is a (P,l)-elation mapping H to any point H' disjoint from $\{P,l\}$ on $PH = h$. The proof of (ii) is similar.

Let us examine more closely the geometrical consequences of the existence in a plane Π of a non-trivial (P,l)-collineation $\mathscr{T}$ with given centre P and axis l. Choose a triangle XYZ whose vertices are not on l and such that $PXYZ$ is a regular quadrangle. Let $\mathscr{T}$ be fixed by the assignment of a point X' such that

$$\mathscr{T}X = X'.$$

Construct $Y' = \mathscr{T}Y, Z' = \mathscr{T}Z$. Then $P = XX' \cap YY' \cap ZZ'$, that is, the triangles XYZ and $X'Y'Z'$ are in *central perspective* from the point P. Moreover, $XY \cap X'Y'$, $YZ \cap Y'Z'$, $ZX \cap Z'X'$ are fixed points of $\mathscr{T}$ and distinct from P (since $PXYZ$ is a regular quadrangle), so that the three points lie on l, or, in other words, the triangles XYZ and $X'Y'Z'$ are in *axial perspective* about the line l.

Given P, l, X and X', and having chosen Y and Z, we may construct

$$L = YZ \cap l, \quad M = ZX \cap l, \quad N = XY \cap l.$$

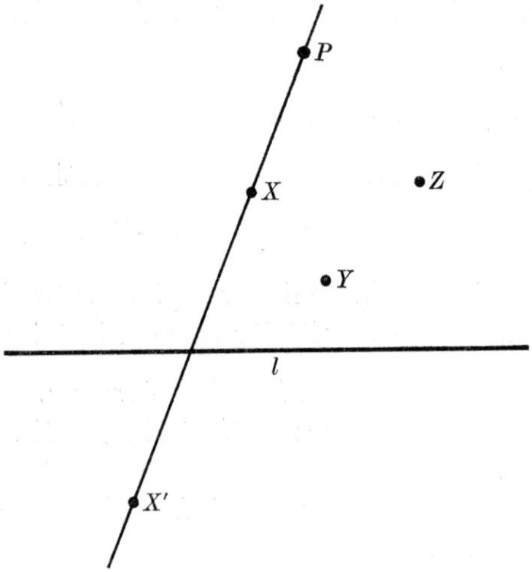

Diagram 2.2 (d)

Then $$Y' = NX' \cap PY \quad \text{and} \quad Z' = MX' \cap PZ.$$

The geometrical consequence of the existence of the central collineation $\mathscr{T}$ is that if we choose Y and Z and construct L, M, N, Y', Z' as above then L, Y', Z' are collinear.

The figure we have constructed is the *Desargues configuration* (G. Desargues, 1593–1662). Although we have shown P and l as non-incident in the diagrams, the argument is valid whether P does or does not lie on l. There may also be further incidences among the points and lines (cf. Room, 1967, p. 31).

Essentially the configuration consists of ten points which are collinear in ten sets of three. The collinearity of the tenth set, L, Y', Z' in the construction above, cannot be deduced from the other collinearities using only the three projective plane axioms.

We can recognize five levels, at least, at which the Desargues configuration could appear in a projective plane Π:

(i) There may be some sets of ten points

$$\{P;\ X, Y, Z;\ X', Y', Z';\ L, M, N\}$$

which form Desargues configurations, but no central collineations, in Π.

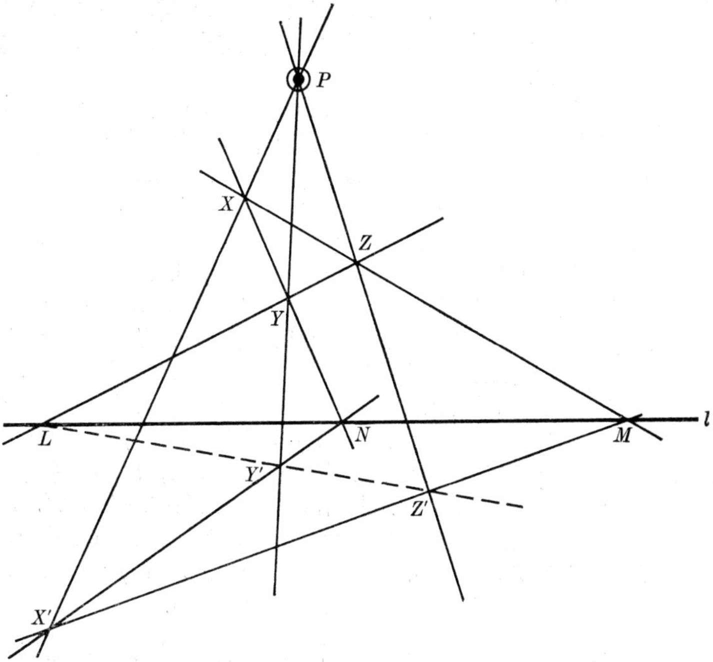

Diagram 2.2 (*e*) The Desargues configuration

(ii) For some choices of P and l, there may be Desargues configurations (with centre P, axis l) sufficient to ensure the existence of at least one (P, l)-collineation.

(iii) For a particular choice of P and l, and for every appropriate choice of X, X', Y, Z, the points L, Y', Z' are collinear. Π is then (P, l)-transitive.

(iv) For a particular choice of l and for every choice of P on l and every appropriate choice of X, X', Y, Z, the points L, Y', Z' are collinear. That is, Π is (P, l)-transitive (all the collineations concerned being elations) for every point P on l. Such a plane is called a *translation plane* with respect to l.

(v) The Desargues condition is satisfied for all allowable choices of P, l, X, X', Y, Z. The plane is *Desarguesian*. All other types of projective planes are termed *non-Desarguesian*.

Of the planes of order 9 discussed in this book, Φ (Chapter 3) is Desarguesian, Ω (Chapter 4) is a translation plane and Ω^D is a 'dual translation plane', while Ψ (Chapter 5), although it possesses central collineations, is not (P, l)-transitive for any choice of P and l.

Whilst it is often possible to verify that a given plane is (P, l)-transitive without invoking the geometrical Desargues condition, usually the most direct way of proving that a plane is not (P, l)-transitive is to construct the first nine sets of three collinear points in a Desargues figure, and verify that the tenth set is not collinear. (See, for example, Theorem 5.5.2.)

Returning to case (iii), let us suppose that Desargues configurations with centre P and axis l result, in Π, from *all* appropriate choices of X, X', Y, Z. Let $\{A, A'\}$ be a pair of points collinear with P and disjoint from $\{P, l\}$. We construct a (P, l)-collineation $\mathscr{T}$ which maps A to A' as follows. $\mathscr{T}(P) = P$ and $\mathscr{T}(L) = L$ for all $L \in l$. If Y is any point not on PA and not on l, $Y' = \mathscr{T}(Y)$ is defined by the construction: $AY \cap l = N$, $PY \cap NA' = Y'$. To define $\mathscr{T}$ on the points of PA, choose a point B not on PA and not on l, and let $B' = \mathscr{T}(B)$. Then the pair B, B' may be used in place of the pair A, A' in the above construction.

EXERCISE 2.2.4 Show that $\mathscr{T}$ is a (P, l)-collineation which maps A to A'. (Cf. M. Hall, 1959, p. 352.)

It follows that Π is (P, l)-transitive in case (iii).

2.3 The plane over an arbitrary field

Let $\mathfrak{K}$ be any field, not necessarily commutative. The non-zero vectors $\mathbf{x} = (x, y, z)$ with components in $\mathfrak{K}$ may be used as (homogeneous) coordinate-vectors for the points of a projective plane, the 'plane over $\mathfrak{K}$', if we introduce the convention that, for any non-zero h in $\mathfrak{K}$, the vectors $\mathbf{x}$ and $\mathbf{x}h$ represent the same point. We denote the plane over $\mathfrak{K}$ by $\Pi(\mathfrak{K})$. A line is the set of points whose coordinate-vectors $\mathbf{x} = (x, y, z)$ satisfy a linear equation $ax + by + cz = 0$ in which the coefficients belong to $\mathfrak{K}$ and are not all three zero. The equation

$$(ka)\,x + (kb)\,y + (kc)\,z = 0 \quad (k \neq 0)$$

represents the same line.

We shall use column-vectors

$$\mathbf{x} = \begin{bmatrix} x \\ y \\ z \end{bmatrix}$$

for points, but shall usually print these as (x, y, z). We shall assign to the line $ax+by+cz = 0$ the row-vector $\mathbf{a}^T = [a, b, c]$ and print the linear expression $ax+by+cz$ as $\mathbf{a}^T\mathbf{x}$ or $[a, b, c]\,(x, y, z)$. The row-vector $k\mathbf{a}^T, k \in \mathfrak{K}$, $k \neq 0$, represents the same line. The point $\mathbf{x}$ lies on the line $\mathbf{a}^T$ if and only if $\mathbf{a}^T\mathbf{x} = 0$. This is equivalent to $k\mathbf{a}^T\mathbf{x}h = 0$ if $k \neq 0$, $h \neq 0$, so the incidence condition is independent of the choice of coordinate-vectors for the point and line.

Let us verify that the system is a projective plane. Distinct points P_1, P_2 given by $\mathbf{x}_1 = (x_1, y_1, z_1)$, $\mathbf{x}_2 = (x_2, y_2, z_2)$ lie on the line $[a, b, c]$ if and only if

$$ax_1+by_1+cz_1 = 0,$$

$$ax_2+by_2+cz_2 = 0.$$

Suppose first that $x_1, y_1, z_1, x_2, y_2, z_2$ are all non-zero. We must be careful not to assume the commutative law for multiplication. Now at least one of $x_1y_1^{-1}-x_2y_2^{-1}$, $y_1z_1^{-1}-y_2z_2^{-1}$, $z_1x_1^{-1}-z_2x_2^{-1}$ is non-zero, since if all are zero then

$$(x_2, y_2, z_2) = (x_1, y_1, z_1)\,h \quad \text{with} \quad h = x_1^{-1}x_2 = y_1^{-1}y_2 = z_1^{-1}z_2,$$

contradicting the distinctness of P_1 and P_2. Without loss of generality, we may suppose that $x_1y_1^{-1}-x_2y_2^{-1} \neq 0$. Dividing the original equations by y_1 and y_2 respectively,

$$ax_1y_1^{-1}+b+cz_1y_1^{-1} = 0,$$

$$ax_2y_2^{-1}+b+cz_2y_2^{-1} = 0,$$

and subtracting, we obtain

$$a(x_1y_1^{-1}-x_2y_2^{-1})+c(z_1y_1^{-1}-z_2y_2^{-1}) = 0.$$

If $z_1y_1^{-1} \neq z_2y_2^{-1}$ then the value of a may be assigned arbitrarily and will determine uniquely a value for c. One of the earlier equations may then be used to determine (uniquely) the value of b. Replacement of a by ka necessitates the replacement of b and c by kb and kc. So there is a non-zero solution $[a, b, c]$ and every solution is a left-multiple of this; that is, P_1 and P_2 determine exactly one line. If $z_1 y_1^{-1} - z_2 y_2^{-1} = 0$, then $a = 0$, but c may be assigned an arbitrary value. The value of b is then determined, and once again all solutions are left-multiples of some non-zero solution.

EXERCISE 2.3.1 Consider the case where one or more of $x_1, y_1, z_1, x_2, y_2, z_2$ is zero. Show that P_1 and P_2 still determine a unique line.

The second axiom for a projective plane may be dealt with similarly. Distinct lines $[a_1, b_1, c_1]$ and $[a_2, b_2, c_2]$ both pass through a point (x, y, z) if and only if

$$a_1 x + b_1 y + c_1 z = 0,$$
$$a_2 x + b_2 y + c_2 z = 0.$$

The solution of this pair of equations is exactly analogous to the solution of the pair considered above.

To verify the third axiom it suffices to consider the four points $(1, 0, 0)$, $(0, 1, 0)$, $(0, 0, 1)$, $(1, 1, 1)$, no three of which are collinear.

THEOREM 2.3.1 *Every field $\mathfrak{K}$ determines a projective plane* $\Pi(\mathfrak{K})$.

Suppose $\mathbf{a}$ and $\mathbf{b}$ are distinct points of $\Pi(\mathfrak{K})$. Every non-zero right linear combination $\mathbf{a}\lambda + \mathbf{b}\mu$ of $\mathbf{a}$ and $\mathbf{b}$ satisfies the equation of the line $\langle \mathbf{a}, \mathbf{b} \rangle$, and therefore represents a point on that line.

EXERCISE 2.3.2 Show that every point on $\langle \mathbf{a}, \mathbf{b} \rangle$ is a right linear combination of $\mathbf{a}$ and $\mathbf{b}$.

THEOREM 2.3.2 *The coordinate-vectors of the vertices of any regular quadrangle in* $\Pi(\mathfrak{K})$ *may be selected so that their sum is the zero vector.*

Let $\mathbf{a}, \mathbf{b}, \mathbf{c}, \mathbf{d}$ be the vertices of a regular quadrangle. If $\mathfrak{K}$ is a commutative field we may say at once that, because the vectors have only three components, they are linearly dependent, that is, there are multipliers, $\kappa, \rho, \sigma, \tau$ not all zero, such that

$$\mathbf{a}\rho + \mathbf{b}\sigma + \mathbf{c}\tau + \mathbf{d}\kappa = \mathbf{0}.$$

In the skew field the result is still valid. We take

$$\mathbf{e}_1 = (1, 0, 0), \quad \mathbf{e}_2 = (0, 1, 0), \quad \mathbf{e}_3 = (0, 0, 1),$$

so that $\mathbf{a} = \Sigma\, \mathbf{e}_i a_i$, etc. Since the lines $a_2x + b_2y + c_2z = 0$ and $a_3x + b_3y + c_3z = 0$ have a common point, multipliers (not all zero) can be found such that

$$\mathbf{a}\lambda_1 + \mathbf{b}\mu_1 + \mathbf{c}\nu_1 = \mathbf{e}_1\theta_1.$$

Thus for each $i = 1, 2, 3$, four multipliers (not all zero) can be found such that

$$\mathbf{a}\lambda_i + \mathbf{b}\mu_i + \mathbf{c}\nu_i = \mathbf{e}_i\theta_i.$$

Since $\mathbf{d} = \Sigma \mathbf{e}_i d_i$, the result follows provided each θ_i is non-zero.

Now $\theta_i \neq 0$ because, if $\theta_i = 0$ then one of $\mathbf{a}, \mathbf{b}, \mathbf{c}$ is a right linear combination of the other two, contradicting the hypothesis that $\mathbf{a}, \mathbf{b}, \mathbf{c}, \mathbf{d}$ is a regular quadrangle.

By absorbing the multipliers into the vectors we may re-write the relation $\mathbf{a}\rho + \mathbf{b}\sigma + \mathbf{c}\tau + \mathbf{d}\kappa = \mathbf{0}$ as $\mathbf{a} + \mathbf{b} + \mathbf{c} + \mathbf{d} = \mathbf{0}$.

We remark that if $\mathbf{a}\rho + \mathbf{b}\sigma + \mathbf{c}\tau + \mathbf{d}\kappa = 0$, the point common to the lines $\langle \mathbf{a}, \mathbf{b}\rangle$ and $\langle \mathbf{c}, \mathbf{d}\rangle$ is $\mathbf{a}\rho + \mathbf{b}\sigma$ (which is the same point as $\mathbf{c}\tau + \mathbf{d}\kappa$).

THEOREM 2.3.3 THE PAPPUS (OR PASCAL) CONDITION† *Given $\{L, M, N\}$ and $\{L', M', N'\}$, two sets of three distinct collinear points on distinct lines in $\Pi(\mathfrak{K})$, and such that none of the six points lies at the intersection of the two lines, construct $P = MN' \cap M'N$, $Q = NL' \cap N'L$, $R = LM' \cap L'M$. Then P, Q, R are collinear for all choices of the two sets if and only if $\mathfrak{K}$ is a commutative field.*

The theorem may be proved in the sequence of steps set out in the following exercise:

EXERCISE 2.3.3 (i) Verify that $LNM'Q$ is a regular quad-rangle.

† The first record of this condition as a theorem in Euclidean geometry is in a manuscript attributed to Pappus (c. A.D. 300). The name Pascal is attached because the theorem is a special case of that proved by Blaise Pascal (1623–62) for conics. A plane which satisfies the Pascal condition for all relevant sets of points is referred to as 'Pascalian', and it can be proved (e.g. Room, 1967, p. 241) that every Pascalian plane can be coordinatized by a commutative field.

(ii) Take $L = \mathbf{a}, N = \mathbf{b}, Q = \mathbf{c}, M' = \mathbf{d}$, with

$$\mathbf{a}+\mathbf{b}+\mathbf{c}+\mathbf{d} = \mathbf{0}; \quad \text{and} \quad L' = \mathbf{b}+\mathbf{c}\lambda, \quad N' = \mathbf{a}+\mathbf{c}\nu.$$

Prove that $\lambda+\nu = 1$.

(iii) Take $M = \mathbf{a}+\mathbf{b}\mu$, then

$$R = (\mathbf{b}+\mathbf{c}\lambda)+(\mathbf{a}+\mathbf{b}\mu)\rho = \mathbf{a}\sigma+\mathbf{d}\tau$$

for some ρ, σ, τ; show that R is $\mathbf{a}-(\mathbf{b}+\mathbf{c})\kappa\mu$, where $\kappa = \lambda\nu^{-1}$.

(iv) $P = (\mathbf{a}+\mathbf{b}\mu)\rho'+(\mathbf{a}+\mathbf{c}\nu) = \mathbf{b}\sigma'+\mathbf{d}\tau'$ for some ρ', σ', τ'; show that P is $\mathbf{a}-\mathbf{b}\mu\kappa+\mathbf{c}$.

(v) Prove that P, Q, R are collinear if and only if $\mu\kappa = \kappa\mu$.

For the remainder of this chapter we restrict our discussion to the case where the field $\mathfrak{K}$ is commutative, and use the notation:

$$\mathscr{K} = \text{an arbitrary commutative field,}$$

$$\Pi(\mathscr{K}) = \text{the plane over } \mathscr{K}.$$

2.4 The projectivities of $\Pi(\mathscr{K})$

The central collineations of a plane generate a group, the *group of projectivities*, which is a subgroup of the full collineation group. A projectivity or *projective collineation* is thus, by definition, any collineation which may be written as the product, in some order, of a finite number of central collineations: $\mathscr{T}_1\mathscr{T}_2\ldots\mathscr{T}_r$. (We need not consider powers, as both the positive and negative powers of a central collineation are again central.)

We are to determine the group of projectivities for $\Pi(\mathscr{K})$, the plane over an arbitrary commutative field. Before doing so, we mention a useful criterion for the collinearity of three points in $\Pi(\mathscr{K})$. Recall that, if $\mathbf{x}_1$ and $\mathbf{x}_2$ are distinct points, then $\mathbf{x}_3$ lies on $\langle\mathbf{x}_1, \mathbf{x}_2\rangle$ if and only if $\mathbf{x}_3 = \mathbf{x}_1 h+\mathbf{x}_2 k$ for some $h, k \in \mathscr{K}$ (and not both zero). It follows that three points $\mathbf{x}_1, \mathbf{x}_2, \mathbf{x}_3$ are collinear if and only if there is a non-zero column-vector $\mathbf{u} = (u, v, w)$ such that

$$\mathbf{x}_1 u+\mathbf{x}_2 v+\mathbf{x}_3 w = \mathbf{0},$$

that is,

$$[\mathbf{x}_1, \mathbf{x}_2, \mathbf{x}_3]\,\mathbf{u} = \mathbf{0}.$$

We deduce:

THEOREM 2.4.1 *Three points* $\mathbf{x}_1$, $\mathbf{x}_2$, $\mathbf{x}_3$ *in* $\Pi(\mathscr{K})$ *are collinear if and only if the matrix* $[\mathbf{x}_1, \mathbf{x}_2, \mathbf{x}_3]$, *whose columns are the coordinate-vectors of the points, is singular.*

Some collineations of $\Pi(\mathscr{K})$ are easily identified, for let

$$\mathbf{A} = \begin{bmatrix} a_1 & b_1 & c_1 \\ a_2 & b_2 & c_2 \\ a_3 & b_3 & c_3 \end{bmatrix}$$

be any 3×3 non-singular matrix with components in $\mathscr{K}$, and let $\mathbf{x} = (x, y, z)$, a column-vector, be the coordinate-vector of a variable point. Then $\mathbf{x} \to \mathbf{Ax}$ gives a 1–1 transformation of the set of points onto itself (the inverse transformation is given by $\mathbf{x} \to \mathbf{A}^{-1}\mathbf{x}$); the line

$$\mathbf{u}^T\mathbf{x} \equiv (\mathbf{u}^T\mathbf{A}^{-1})(\mathbf{Ax}) = 0$$

is mapped to the line $\mathbf{u}^T\mathbf{A}^{-1}\mathbf{x} = 0$; that is, $((\mathbf{A}^{-1})^T\mathbf{u})^T\mathbf{x} = 0$. It is to be proved later that this collineation is a projectivity.

EXERCISE 2.4.1 Show that in the plane $\Delta = \Pi(\mathscr{D})$, where $\mathscr{D}$ is the Galois field of order 3, the collineations $\mathscr{A}_i$ determined by the matrices

$$\mathscr{A}_1\colon \begin{bmatrix} 0 & 0 & 1 \\ 1 & 0 & 1 \\ 0 & 1 & 0 \end{bmatrix}, \quad \mathscr{A}_2\colon \begin{bmatrix} 0 & 0 & 1 \\ 1 & 0 & 1 \\ 0 & 1 & -1 \end{bmatrix}, \quad \mathscr{A}_3\colon \begin{bmatrix} 0 & 0 & -1 \\ 1 & 0 & -1 \\ 0 & 1 & 0 \end{bmatrix},$$

are such that: $\mathscr{A}_1$ permutes the points of Δ in a single cycle, $\mathscr{A}_2$ fixes two points and permutes the others in cycles of 2, 3 and 6 points, $\mathscr{A}_3$ fixes one point, permutes a set of 4 collinear points and permutes the remaining points as a cycle of 8.

Let us examine the collineation $\mathscr{A}$ of the plane $\Pi(\mathscr{K})$ corresponding to a matrix $\mathbf{A}$ of particularly simple form. Take

$$\mathbf{A} = \begin{bmatrix} a & 0 & 0 \\ 0 & 1 & 0 \\ 0 & 0 & 1 \end{bmatrix}$$

with $a \neq 0, 1$ (if $a = 0$, $\mathbf{A}$ is singular; if $a = 1$ then $\mathscr{A} = 1$, that is, $\mathscr{A}$ maps each point to itself). Then $X = (x, y, z) \to (ax, y, z)$, so that $P_1 = (1, 0, 0)$ is fixed by $\mathscr{A}$ and every point on the line $l_1 = [1, 0, 0]$ is fixed. Now, for any x, y, z,

$$\begin{vmatrix} 1 & x & ax \\ 0 & y & y \\ 0 & z & z \end{vmatrix} = 0,$$

and therefore $P_1, X, \mathscr{A}X$ are collinear; that is, P_1 is a centre for $\mathscr{A}$. Thus $\mathscr{A}$ is a (P_1, l_1)-collineation.

EXERCISE 2.4.2 (i) Show that $\Pi(\mathscr{K})$ is (P_1, l_1)-transitive.

(ii) Let $l_2 = [0, 1, 0]$. Find matrices which represent the (P_1, l_2)-collineations of $\Pi(\mathscr{K})$ and show that $\Pi(\mathscr{K})$ is (P_1, l_2)-transitive.

(iii) P, l is any point-line pair in $\Pi(\mathscr{K})$. Prove that there is, if $P \notin l$, a collineation which maps P, l to P_1, l_1, or if $P \in l$, a collineation which maps P, l to P_1, l_2.

This exercise and Theorem 2.2.3 together imply:

THEOREM 2.4.2 *$\Pi(\mathscr{K})$ is (P, l)-transitive for all point-line pairs P, l; that is, $\Pi(\mathscr{K})$ is Desarguesian.*

We give now an explicit matrix representation of an arbitrary central collineation in terms of its centre, its axis, one point and the transform of that point (cf. Theorem 2.2.1).

THEOREM 2.4.3 *The (P, l)-collineation $\mathscr{T}$ which maps*

H $(H \neq P, H \notin l)$ to H' has matrix $(\mathbf{a}^T\mathbf{h})\,\mathbf{1} + c\mathbf{p}\mathbf{a}^T$, *where*

$$\mathbf{p} = P, \quad \mathbf{a}^T = l, \quad \mathbf{h} = H, \quad c\mathbf{p} + \mathbf{h} = H'.$$

Let $X = \mathbf{x}$ be any point not on l or PH, and let $\mathscr{T}X = X' = \mathbf{x}'$; also let $\mathbf{k} = K = HX \cap l$. Then $\mathbf{a}^T\mathbf{k} = 0$ and, for some $r, s \in \mathscr{K}$, $\mathbf{k} = r\mathbf{x} + s\mathbf{h}$. It follows that $r(\mathbf{a}^T\mathbf{x}) = -s(\mathbf{a}^T\mathbf{h})$, and we may therefore take $\mathbf{k} = (\mathbf{a}^T\mathbf{h})\,\mathbf{x} - (\mathbf{a}^T\mathbf{x})\,\mathbf{h}$. Now $X' = PX \cap H'K$; the point $(\mathbf{a}^T\mathbf{h})\,\mathbf{x} + (c\mathbf{a}^T\mathbf{x})\,\mathbf{p}$ lies on PX, and is the same point as

$$\{(\mathbf{a}^T\mathbf{h})\,\mathbf{x} - (\mathbf{a}^T\mathbf{x})\,\mathbf{h}\} + (\mathbf{a}^T\mathbf{x})\,(c\mathbf{p} + \mathbf{h}),$$

which lies on $H'K$. So that

$$\mathbf{x}' = (\mathbf{a}^T\mathbf{h})\,\mathbf{x} + (c\mathbf{a}^T\mathbf{x})\,\mathbf{p} = \{(\mathbf{a}^T\mathbf{h})\,\mathbf{1} + c\mathbf{p}\mathbf{a}^T\}\,\mathbf{x}.$$

This formula may be verified similarly for $X \in PH$. It holds trivially when $X \in l$; that is, $\mathbf{a}^T\mathbf{x} = 0$ and $X' = X$.

The matrix of this theorem represents a homology or an elation according as

$$\mathbf{a}^T\mathbf{p} \neq 0, \quad \text{or} \quad \mathbf{a}^T\mathbf{p} = 0.$$

EXERCISE 2.4.3 (i) Find simple forms of $\mathbf{p}$ and $\mathbf{a}^T$ such that

$$\mathbf{1}+\mathbf{p}\mathbf{a}^T = \begin{bmatrix} 0 & 1 & 0 \\ 1 & 0 & 0 \\ 0 & 0 & 1 \end{bmatrix}.$$

(ii) Express the collineation of order 3,

$$\begin{bmatrix} 0 & 1 & 0 \\ 0 & 0 & 1 \\ 1 & 0 & 0 \end{bmatrix},$$

as the resultant of two homologies.

Every projectivity of $\Pi(\mathscr{K})$ is a product of central collineations, and may therefore be represented as a product of (necessarily non-singular) matrices. Thus:

THEOREM 2.4.4 *Every projective collineation (projectivity), of the plane* $\Pi(\mathscr{K})$, *may be represented by a* 3×3 *non-singular matrix with components in* $\mathscr{K}$.

We know that every 3×3 non-singular matrix represents a collineation; we are to prove that:

THEOREM 2.4.5 *The collineation of* $\Pi(\mathscr{K})$ *represented by any* 3×3 *non-singular matrix is projective.*

Let the matrix be $\mathbf{A}$, the collineation $\mathscr{A}$. We must show that $\mathscr{A}$ is a product of central collineations. $\mathscr{A}$ maps the points $E_1 = (1,0,0) = \mathbf{e}_1$, $E_2 = (0,1,0) = \mathbf{e}_2$, $E_3 = (0,0,1) = \mathbf{e}_3$,

$$U = (1,1,1) = \mathbf{u} \quad \text{to, say,} \quad E_1', E_2', E_3', U'.$$

We show first that there is a projectivity which also does this.

The plane is (P, l)-transitive for all P, l, so that there is a central collineation which maps E_1 to E_1'. Let it map E_2 to F_2; there is a central collineation, with axis through E_1', which maps F_2 to E_2'. The product of the two central collineations is a projectivity $\mathscr{A}_1$ which maps E_1 to E_1' and E_2 to E_2'. If $E_1' \notin E_1E_2$, $E_2' \notin E_1E_2$, $\mathscr{A}_1$ may be taken to be a central collineation (see Diagram 2.4(*a*)). Let $H_3 = \mathscr{A}_1 E_3$, $H_4 = \mathscr{A}_1 U$. There is a collineation $\mathscr{A}_2$ with axis $E_1'E_2'$ and centre collinear with H_3, E_3' which maps H_3

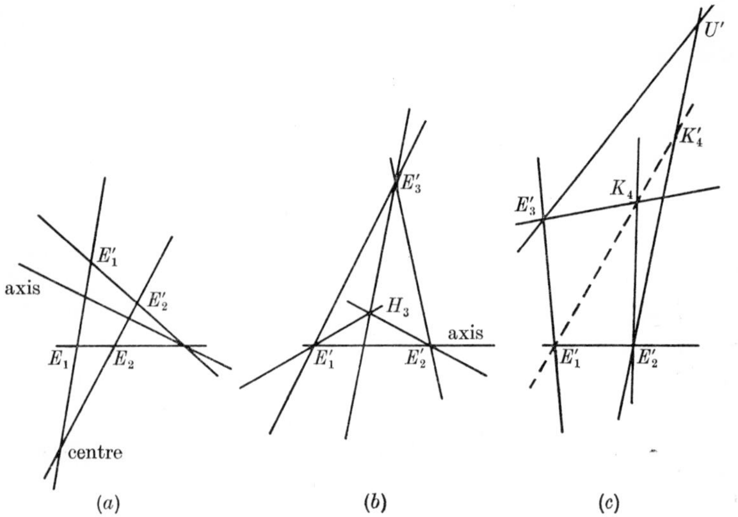

Diagram 2.4

to E_3' (see Diagram 2.4(*b*)). $\mathscr{A}_2$ fixes E_1' and E_2'. Let $K_4 = \mathscr{A}_2 H_4$. There is an $(E_1', E_2'E_3')$-collineation $\mathscr{A}_3$ which maps K_4 to

$$K_4' = E_1'K_4 \cap E_2'U';$$

and an $(E_2', E_1'E_3')$-collineation $\mathscr{A}_4$ which maps K_4' to U' (Diagram 2.4(*c*)). $\mathscr{B} = \mathscr{A}_4\mathscr{A}_3\mathscr{A}_2\mathscr{A}_1$ is the required projectivity.

Now $\mathscr{B}$, being a projectivity, can be represented by a matrix $\mathbf{B}$ (Theorem 2.4.4); also $\mathscr{B}^{-1}$ maps E_1', E_2', E_3', U' to E_1, E_2, E_3, U. So $\mathscr{B}^{-1}\mathscr{A}$ is a collineation represented by a matrix $\mathbf{M} = \mathbf{B}^{-1}\mathbf{A}$ and it fixes E_1, E_2, E_3, U. Let $\mathbf{M} = [\boldsymbol{l}, \mathbf{m}, \mathbf{n}]$. Then there exist non-zero members a, b, c, d of $\mathscr{K}$, such that $[\boldsymbol{l}, \mathbf{m}, \mathbf{n}]\mathbf{e}_1 = a\mathbf{e}_1$, $[\boldsymbol{l}, \mathbf{m}, \mathbf{n}]\mathbf{e}_2 = b\mathbf{e}_2$, $[\boldsymbol{l}, \mathbf{m}, \mathbf{n}]\mathbf{e}_3 = c\mathbf{e}_3$ and $[\boldsymbol{l}, \mathbf{m}, \mathbf{n}]\mathbf{u} = d\mathbf{u}$; that is,

$\boldsymbol{l} = a\mathbf{e}_1$, $\mathbf{m} = b\mathbf{e}_2$, $\mathbf{n} = c\mathbf{e}_3$, $\boldsymbol{l}+\mathbf{m}+\mathbf{n} = d\mathbf{u}$, so that $a = b = c = d$ and $\mathbf{M} = d\mathbf{1}$. It follows that $\mathscr{B}^{-1}\mathscr{A}$ is given by $\mathbf{x} \to d\mathbf{x}$; that is, $\mathscr{B}^{-1}\mathscr{A} = 1$ and so $\mathscr{A}$ is the projectivity $\mathscr{B}$.

A consequence of this analysis of the structure of projectivities is the following theorem:

THEOREM 2.4.6 *In a plane over a commutative field there is a unique projectivity which transforms the vertices in order of any regular quadrangle into the vertices in order of any other regular quadrangle.*

2.5 Change of coordinates. Similarity

In a plane over a (commutative) field $\mathscr{K}$, a matrix equation

$$k\mathbf{y} = \mathbf{A}\mathbf{x} \quad (\mathbf{A}: 3 \times 3, \text{ non-singular})$$

may be given two different geometrical interpretations, one of which was discussed in the previous section: in a fixed coordinate system it represents a projective collineation mapping the point $\mathbf{x} = (x_1, x_2, x_3)$ to the point $\mathbf{y} = (y_1, y_2, y_3)$. Alternatively we may regard $\mathbf{x}$ and $\mathbf{y}$ as two vectors representing the same point in two different coordinate-systems $\mathscr{X}$ and $\mathscr{Y}$. For let $B_1B_2B_3B_u$ be a regular quadrangle with coordinate-vectors, in $\mathscr{X}$,

$$\mathbf{b}_j = (b_{1j}, b_{2j}, b_{3j}) \quad (j = 1, 2, 3, u).$$

Then since B_1, B_2, B_3 are non-collinear, $\mathbf{b}_u$ is a linear combination, say $h_1\mathbf{b}_1+h_2\mathbf{b}_2+h_3\mathbf{b}_3$, of the other three vectors. Because of the non-collinearity conditions, the coefficients h_j are non-zero. The coordinate-vectors could therefore have been chosen so that $\mathbf{b}_u = \mathbf{b}_1+\mathbf{b}_2+\mathbf{b}_3$. Using this convention, the points B_j and B_u determine a system $\mathscr{Y}$ of coordinates; a point P which has coordinate-vector $\mathbf{x} = y_1\mathbf{b}_1+y_2\mathbf{b}_2+y_3\mathbf{b}_3$ in $\mathscr{X}$ is assigned the coordinate-vector $\mathbf{y} = (y_1, y_2, y_3)$ in $\mathscr{Y}$. $B_1B_2B_3$ is called the *triangle of reference* for $\mathscr{Y}$, $B_1B_2B_3B_u$ the *basis*. B_1 has coordinate-vector $(1, 0, 0)$ in $\mathscr{Y}$. If $\mathbf{B} = [\mathbf{b}_1, \mathbf{b}_2, \mathbf{b}_3]$, then $\mathbf{x} = \mathbf{B}\mathbf{y}$. Since the coordinate-vectors are to be unique only to within multiplication by a non-zero scalar, we write

$$k\mathbf{x} = \mathbf{B}\mathbf{y},$$

this equation being interpreted thus: if $\mathbf{x}$ represents a point P in $\mathscr{X}$ and $\mathbf{y}$ represents P in $\mathscr{Y}$ (where $\mathscr{Y}$ is determined by points B_j, B_u having coordinate-vectors $\mathbf{b}_j, \mathbf{b}_u = \mathbf{b}_1+\mathbf{b}_2+\mathbf{b}_3$ in $\mathscr{X}$), then, for some $k \neq 0$, $k\mathbf{x} = \mathbf{B}\mathbf{y}$. To obtain the alternative interpretation of the equation $k\mathbf{y} = \mathbf{A}\mathbf{x}$, take $\mathbf{B} = \mathbf{A}^{-1}$ and replace k by $1/k$.

Let $h\mathbf{x}' = \mathbf{H}\mathbf{x}$ be the equation in $\mathscr{X}$ of a projectivity, and let the coordinate-vectors in $\mathscr{Y}$ of the points $\mathbf{x}, \mathbf{x}'$ be determined by $k\mathbf{y} = \mathbf{A}\mathbf{x}$, $k'\mathbf{y}' = \mathbf{A}\mathbf{x}'$. Then $hk'\mathbf{y}' = h\mathbf{A}\mathbf{x}' = \mathbf{A}\mathbf{H}\mathbf{x} = k\mathbf{A}\mathbf{H}\mathbf{A}^{-1}\mathbf{y}$. Thus:

THEOREM 2.5.1 *If, in the plane* $\Pi(\mathscr{K})$, *the coordinate-vectors* $\mathbf{x}$, $\mathbf{y}$ *of a point in two different systems are related by a matrix equation* $k\mathbf{y} = \mathbf{A}\mathbf{x}$, *then the projectivity* $\mathbf{x} \to \mathbf{H}\mathbf{x}$ *becomes* $\mathbf{y} \to \mathbf{K}\mathbf{y}$ *where* $\mathbf{K} = \mathbf{A}\mathbf{H}\mathbf{A}^{-1}$.

DEFINITION 2.5.1 Matrices $\mathbf{H}$, $\mathbf{K}$ for which there exists a non-singular matrix $\mathbf{B}$ such that $\mathbf{K} = \mathbf{B}\mathbf{H}\mathbf{B}^{-1}$ are said to be *similar*.†

EXERCISE 2.5.1 Show that similarity is an equivalence relation (reflexive, symmetric, transitive).

The equivalence class of matrices similar to a non-singular matrix $\mathbf{H}$ is a set of matrices which represent, in the various coordinate-systems, the projectivity determined by $\mathbf{H}$ in some given coordinate-system. For a fixed coordinate-system the matrices representing projectivities form a group under multiplication (which is the operation corresponding to the composition of the projectivities). This group is of course unaffected by changes of coordinates.

The set of matrices similar to a given matrix $\mathbf{H}$ has some remarkable properties. One of very considerable significance in the work which follows is:

THEOREM 2.5.2 *If* $\mathbf{K}$ *is similar to* $\mathbf{H}$ *then the polynomials*

$$\det(x\mathbf{1}-\mathbf{H}) \quad \textit{and} \quad \det(x\mathbf{1}-\mathbf{K})$$

are identical.

† It should be noticed that for any two matrices $\mathbf{H}$, $\mathbf{K}$ there exist matrices $\mathbf{F}$ such that $\mathbf{KF} = \mathbf{FH}$. The essential property of similarity is that there exist *non-singular* matrices $\mathbf{F}$.

$$\mathbf{K} = \mathbf{AHA}^{-1} \Rightarrow x\mathbf{1}-\mathbf{K} = x\mathbf{1}-\mathbf{AHA}^{-1} = \mathbf{A}(x\mathbf{1}-\mathbf{H})\,\mathbf{A}^{-1}$$
$$\Rightarrow \det(x\mathbf{1}-\mathbf{K}) = \det[\mathbf{A}(x\mathbf{1}-\mathbf{H})\mathbf{A}^{-1}]$$
$$= \det(x\mathbf{1}-\mathbf{H}).$$

DEFINITION 2.5.2 The polynomial $\det(x\mathbf{1}-\mathbf{H})$ in x over $\mathscr{K}$ is called the *characteristic polynomial* or *characteristic function* of $\mathbf{H}$. If $\det(a\mathbf{1}-\mathbf{H}) = 0$; that is, if a is a zero of the characteristic polynomial, then a is called an *eigenvalue*† of $\mathbf{H}$. If a is an eigenvalue then there is a non-zero vector $\mathbf{u}$ such that

$$(a\mathbf{1}-\mathbf{H})\,\mathbf{u} = \mathbf{0},$$

that is, $\mathbf{Hu} = a\mathbf{u}$; every such vector $\mathbf{u}$ is called an *eigenvector* of $\mathbf{H}$ (corresponding to the eigenvalue a).

If the characteristic polynomial is irreducible over the field $\mathscr{K}$ then $\mathbf{H}$ has no eigenvalue. In §3.2 and §3.6 we shall be particularly interested in matrices with irreducible characteristic polynomials, and the following theorem will play an essential role:

THEOREM 2.5.3 *If two* 3×3 *matrices* $\mathbf{H}, \mathbf{K}$ *over a commutative field* $\mathscr{K}$ *have the same characteristic polynomial, and this polynomial is irreducible in* $\mathscr{K}$, *then* $\mathbf{H}$ *and* $\mathbf{K}$ *are similar.*

As a preliminary to this theorem we prove two lemmas; the first is the Cayley–Hamilton theorem 'every matrix satisfies its own characteristic equation':

LEMMA 1 *If* $\det(x\mathbf{1}-\mathbf{H}) \equiv \chi(x)$, *then* $\chi(\mathbf{H}) = \mathbf{0}$.

Let

(i) $\chi(x) \equiv x^3+lx^2+mx+n.$

Consider the adjoint matrix, $\mathbf{A}$, of $x\mathbf{1}-\mathbf{H}$, namely, the matrix whose elements are the cofactors of the corresponding elements of the transposed matrix $(x\mathbf{1}-\mathbf{H})^T$. Each of these cofactors is a 2×2 determinant whose value is a polynomial of degree less than 3 in x, so that there exist matrices $\mathbf{P}, \mathbf{Q}, \mathbf{R}$ such that:

(ii) $\mathbf{A} = x^2\mathbf{P}+x\mathbf{Q}+\mathbf{R}.$

† 'Eigen' (German) = 'own'—a value which 'belongs to the matrix'.

Also, from direct matrix multiplication,

$$\text{(iii)} \qquad \mathbf{A}(x\mathbf{1}-\mathbf{H}) = \det(x\mathbf{1}-\mathbf{H})\,\mathbf{1}.$$

From relations (i), (ii) and (iii) we deduce that the relation

$$(x^2\mathbf{P}+x\mathbf{Q}+\mathbf{R})\,(x\mathbf{1}-\mathbf{H}) = (x^3+lx^2+mx+n)\,\mathbf{1}$$

holds. It follows that

$$\begin{aligned} \mathbf{P} &= \mathbf{1}, \\ \mathbf{Q}-\mathbf{PH} &= l\mathbf{1}, \\ \mathbf{R}-\mathbf{QH} &= m\mathbf{1}, \\ -\mathbf{RH} &= n\mathbf{1}, \end{aligned}$$

and therefore that

$$\begin{aligned} \mathbf{0} &= \mathbf{PH}^3+(\mathbf{Q}-\mathbf{PH})\,\mathbf{H}^2+(\mathbf{R}-\mathbf{QH})\,\mathbf{H}-\mathbf{RH} \\ &= \mathbf{H}^3+l\mathbf{H}^2+m\mathbf{H}+n\mathbf{1} = \chi(\mathbf{H}). \end{aligned}$$

LEMMA 2 *If the characteristic function,* $\chi(x) \equiv x^3+lx^2+mx+n$, *of a matrix* $\mathbf{H}$ *is irreducible, then for every vector* $\mathbf{v}$ (*other than* $\mathbf{v} = \mathbf{0}$) *the matrix*

$$\mathbf{V} \equiv [\mathbf{v}, \mathbf{Hv}, \mathbf{H}^2\mathbf{v}]$$

is non-singular.

Since we cannot have $n = 0$, $\mathbf{H}$ is non-singular. Assume that for some non-zero vector, $\mathbf{v}$, the matrix $\mathbf{V}$ is singular; then there exist scalars r, s, t not all zero such that:

$$r\mathbf{H}^2\mathbf{v}+s\mathbf{Hv}+t\mathbf{1v} = \mathbf{0}.$$

Consider the two polynomials $\chi(x)$ and $\phi(x) \equiv rx^2+sx+t$; since χ is irreducible and of degree 3, and ϕ is of lower degree, χ and ϕ can have no common factor, and there exist therefore pairs of polynomials $\lambda(x)$, $\mu(x)$ such that

$$\lambda(x)\,\phi(x)+\mu(x)\,\chi(x) = 1.$$

This identity is equivalent to a set of relations among the coefficients of the four polynomials, so that

$$\lambda(\mathbf{H})\,\phi(\mathbf{H})+\mu(\mathbf{H})\,\chi(\mathbf{H}) = \mathbf{1}$$

and $$\lambda(\mathbf{H})\,(\phi(\mathbf{H})\,\mathbf{v})+\mu(\mathbf{H})\,(\chi(\mathbf{H})\,\mathbf{v}) = \mathbf{v} \quad \text{for all} \quad \mathbf{v}.$$

But $\chi(\mathbf{H}) = \mathbf{0}$ (Lemma 1) and we have assumed that $\phi(\mathbf{H})\,\mathbf{v} = \mathbf{0}$, so that we have reached a contradiction. There is consequently no polynomial ϕ and therefore $\mathbf{V}$ is non-singular for all non-zero vectors $\mathbf{v}$.

To prove the main theorem:

take
$$\mathbf{G} = \begin{bmatrix} 0 & 0 & -n \\ 1 & 0 & -m \\ 0 & 1 & -l \end{bmatrix}$$

and, as before,
$$\mathbf{V} = [\mathbf{v}, \mathbf{H}\mathbf{v}, \mathbf{H}^2\mathbf{v}],$$

then
$$\begin{aligned} \mathbf{VG} &= [\mathbf{H}\mathbf{v}, \mathbf{H}^2\mathbf{v}, -(l\mathbf{H}^2 + m\mathbf{H} + n\mathbf{1})\,\mathbf{v}] \\ &= [\mathbf{H}\mathbf{v}, \mathbf{H}^2\mathbf{v}, \mathbf{H}^3\mathbf{v}] \\ &= \mathbf{HV}. \end{aligned}$$

But $\mathbf{V}$ is non-singular (Lemma 2), so that $\mathbf{H} = \mathbf{VGV}^{-1}$. That is, if $\mathbf{H}$ is any matrix with an irreducible characteristic function

$$\chi(x) \equiv x^3 + lx^2 + mx + n,$$

then it is similar to $\mathbf{G}$, and therefore all such matrices are similar.

The value of Theorem 2.5.3 in the investigation of the planes $\Pi(\mathscr{K})$ is that it provides a simple 'canonical form' for a matrix with a given characteristic function, provided that the function is irreducible. Thus if

$$\det(x\mathbf{1} - \mathbf{H}) = x^3 + lx^2 + mx + n$$

and the polynomial is irreducible over $\mathscr{K}$, then the matrix

$$\mathbf{G} = \begin{bmatrix} 0 & 0 & -n \\ 1 & 0 & -m \\ 0 & 1 & -l \end{bmatrix}$$

is similar to $\mathbf{H}$, and by a change of coordinate-system (Theorem 2.5.1) we may replace $\mathbf{H}$ by $\mathbf{G}$.

In Chapter 3 we are to be concerned with powers of a matrix, $\mathbf{G}$, and the relations of their characteristic functions to $\chi(\mathbf{G})$, and shall need the theorems which follow.

Theorem 2.5.4 *If* $\chi(\mathbf{H}) = x^3+lx^2+mx+n$ *and* $\chi(\mathbf{H})$ *is irreducible, then*:

(i) $\chi(\mathbf{H}^2) = x^3+(2m-l^2)\,x^2+(m^2-2ln)\,x-n^2$.

(ii) *Over* $GF(3)$, $\chi(\mathbf{H}^3) = \chi(\mathbf{H})$.

(iii) *Over* $GF(9)$, $\chi(\mathbf{H}^9) = \chi(\mathbf{H})$.

(i) $\mathbf{H}$ is similar to

$$\mathbf{G} = \begin{bmatrix} 0 & 0 & -n \\ 1 & 0 & -m \\ 0 & 1 & -l \end{bmatrix},$$

$$x\mathbf{1}-\mathbf{G}^2 = \begin{bmatrix} x & n & -ln \\ 0 & x+m & n-lm \\ -1 & l & x+m-l^2 \end{bmatrix},$$

giving $\qquad \chi(\mathbf{G}^2) = x^3+(2m-l^2)\,x^2+(m^2-2ln)\,x-n^2.$

Also $\quad \mathbf{H} = \mathbf{AGA}^{-1} \Rightarrow \mathbf{H}^2 = \mathbf{AG}^2\mathbf{A}^{-1} \Rightarrow \chi(\mathbf{H}^2) = \chi(\mathbf{G}^2).$

(ii) Over GF(3) all numerical coefficients, and in particular all binomial coefficients, are reduced modulo 3, so that any binomial coefficient which is a multiple of 3 may be replaced by 0. Also, if $n \in$ GF(3), then $n^3 = n$. Thus

$$(y\mathbf{1}-\mathbf{H})^3 = y^3\mathbf{1}-\mathbf{H}^3$$

and

$$\begin{aligned} \det(y^3\mathbf{1}-\mathbf{H}^3) &= [\det(y\mathbf{1}-\mathbf{H})]^3 \\ &= (y^3+ly^2+my+n)^3 \\ &= y^9+l^3y^6+m^3y^3+n^3, \end{aligned}$$

so that

$$\chi(\mathbf{H}^3) = \det(x\mathbf{1}-\mathbf{H}^3) = x^3+lx^2+mx+n = \chi(\mathbf{H}).$$

(iii) For GF(9), write the elements as $r+s\epsilon$ (see § 1.2), where, again, numerical coefficients are reduced modulo 3, and

$$m \in \mathrm{GF}(9) \Rightarrow m^9 = m.$$

In the expansion of $(x^3+lx^2+mx+n)^9$ we obtain coefficients

$$9!/(r!\,s!\,t!(9-r-s-t)!).$$

Apart from the value 1 for the coefficients of x^{27}, l^9x^{18}, m^9x^9, n^9, every coefficient contains a multiple of 3, because 9! contains a factor 3^4, while, except in the cases noted, the denominator contains no higher power of 3 than 3^3. The method of proof used in part (ii) now carries through.

2.6 Conics and polarities in $\Pi(\mathscr{K})$

Let us consider one of the methods of constructing a parabola $y^2 = x$ in a real Cartesian plane, which does not, at least explicitly, involve any process of measurement, but only the geometric operations of drawing lines parallel to the coordinate axes and through pairs of given points.

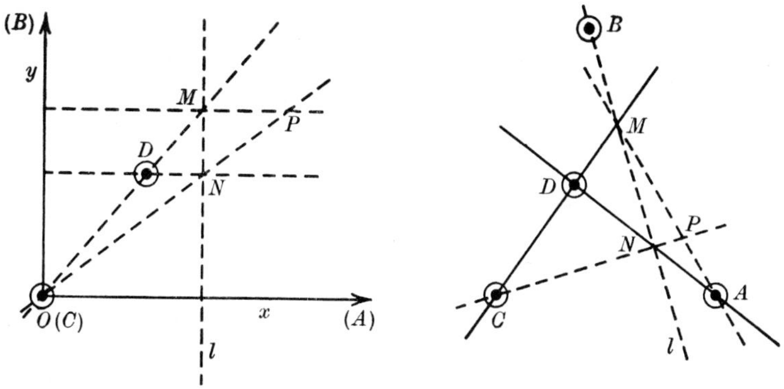

(a) (b)

Diagram 2.6

Let D be any fixed point and l a line of the set parallel to Oy. Construct the points M, N, P as follows: $M = l \cap DO$, $N \in l$ and $ND \| Ox$, $P \in ON$ and $PM \| Ox$. The parabola is defined as the set of points $\{P\}$ derived by this construction from the lines of the set $\{l\}$. Take D to be the point (d, d') and any selected line l to be $x = x_0$, then OD is $dy = d'x$, M is $(x_0, d'x_0/d)$, N is (x_0, d'), ON is $x_0 y = d'x$, and MP is $y = d'x_0/d$, so that

$$P = ON \cap MP \quad \text{is} \quad (x_0^2/d, d'x_0/d).$$

Thus the equation satisfied by the coordinates of the points of the set $\{P\}$ constructed in this way from the lines of the set $\{l\}$ is $y^2 = d'^2 x/d$, and by selecting suitable new scales on the axes we obtain the required equation $y^2 = x$.

Now replace the Cartesian coordinates (x, y) by homogeneous coordinates (x', y', z') by the sequence of substitutions

$$(x, y) \to (x, y, 1) \to (xk, yk, k) = (x', y', z')$$

and then write (x, y, z) for (x', y', z'). We have therefore to adjoin the line $z = 0$ to the Cartesian plane in order to obtain the projective plane over the real field, but apart from points on $z = 0$ there is an exact 1–1 correspondence between points of the two planes. The set of parallel lines $y = mx+c$, m fixed, c variable, in the Cartesian plane is replaced by the set of lines $y = mx+cz$ which are concurrent in the point $(1, m, 0)$ on the line $z = 0$.

Finally, instead of restricting the coordinates to the real field, take them to be members of any commutative field $\mathscr{K}$ which is not of characteristic 2 (that is, $1+1 \neq 0$).

In the construction above, rename O as C, and take $y = 0$, $z = 0$ to be A and $x = 0$, $z = 0$ to be B; that is, $A = (1, 0, 0)$, $B = (0, 1, 0)$, $C = (0, 0, 1)$. Rephrasing the description of the construction in terms of these elements we have:

DEFINITION 2.6.1 *The DTP construction of a conic.*† *ABCD* is a regular quadrangle in $\Pi(\mathscr{K})$, l is an arbitrary line through B. $M = DC \cap l$, $N = DA \cap l$, $P = NC \cap AM$. The *conic* Γ is the set of points $\{P\}$ corresponding to the lines of the set $\{l : l \supset B\}$. (Diagram 2.6 (*b*).)

From the DTP construction we can obtain only a 'non-singular' conic; that is, we cannot construct in this way a conic all of whose points lie on two lines or on one line, or a conic which reduces to a single point (for example, $x^2+y^2 = 0$ in the real Cartesian plane).

Γ contains A, C, D (but not B). If we take D to be $(1, 1, 1)$, then AD is $y = z$, and CD is $x = y$. If l is $rx = tz$, then $M = (t, t, r)$,

† 'DTP' = 'Doubly-tangential Pascal'. In its general form the Pascal construction provides a method of constructing points of the conic which contains a given set of five points. Given $\{A_i\}$, a set of five points no three collinear, and l any line through, say, A_1, then the Pascal construction determines the second point of the conic containing $\{A_i\}$ which lies on l. The DTP construction can be regarded as the special case in which

$$A_2 = A_3 = A, \quad A_4 = A_5 = C, \quad A_1 = D;$$

the conic touches AB at A, CB at C and passes through D.

$N = (t, r, r)$, AM is $ry = tz$, CN is $rx = ty$, and $P = (t^2, tr, r^2)$. Expressed in its simplest form, Γ is the set of points

$$\{(t^2, t, 1) : t \in \mathscr{K}\} \cup \{A\}.$$

The equation of Γ is $xz - y^2 = 0$, which we may write as

$$[x, y, z] \begin{bmatrix} 0 & 0 & 1 \\ 0 & -2 & 0 \\ 1 & 0 & 0 \end{bmatrix} \begin{bmatrix} x \\ y \\ z \end{bmatrix} = 0, \quad \text{say} \quad \mathbf{x}^T\mathbf{C}\mathbf{x} = 0.$$

(It should be noticed that until this last statement is made, the field could have characteristic 2 without in any way modifying the construction or argument.) This form suggests the following relation between row-vectors $[u, v, w]$ and column-vectors (x, y, z):

$$\mathbf{u}^T\mathbf{C}\mathbf{x} \equiv wx - 2vy + uz = 0.$$

From any point (u, v, w) we derive a unique line $[w, -2v, u]$, and from any line $[a, b, c]$ we derive a unique point $(c, -\frac{1}{2}b, a)$; that is, the conic Γ establishes a 1–1 correspondence between the points of the plane and the lines of the plane.

DEFINITION 2.6.2 In relation to the conic $\Gamma: y^2 = xz$, the line $[w, -2v, u]$ is the *polar* of the point (u, v, w) and the point $(w, -\frac{1}{2}v, u)$ is the *pole* of the line $[u, v, w]$. Γ defines a *polarity* $\mathscr{C}$ in the plane $\Pi(\mathscr{K})$, char $\mathscr{K} \neq 2$.

The polarity $\mathscr{C}$ maps the plane onto itself, every point mapping to a line, and every line to a point. If $P = \mathbf{x}$ and $Q = \mathbf{y}$ are points such that $P \in \mathscr{C}Q$, that is $\mathbf{y}^T\mathbf{C}\mathbf{x} = 0$, then $\mathbf{x}^T\mathbf{C}\mathbf{y} = 0$ (as $\mathbf{C}$ is symmetric), and so $Q \in \mathscr{C}P$. Thus:

THEOREM 2.6.1 *The polarity* $\mathscr{C}: (u, v, w) \leftrightarrow wx - 2vy + uz = 0$ *has the property*

$$P \in \mathscr{C}Q \Rightarrow Q \in \mathscr{C}P.$$

Let $P = (t^2, tr, r^2)$ be any point of Γ, l be an arbitrary line through P and $Q = (a, b, c)$, $\neq P$, be some other point on l. Then the points of l on Γ have coordinates $(t^2 + \lambda a, tr + \lambda b, r^2 + \lambda c)$, where λ is such that

$$(t^2 + \lambda a)(r^2 + \lambda c) - (tr + \lambda b)^2 = 0;$$

that is,
$$\lambda(ar^2 - 2btr + ct^2) + \lambda^2(ac - b^2) = 0.$$

The line l therefore contains, in general, P and exactly one other point of Γ, the only exceptional cases being those for which Q is such that $ar^2 - 2btr + ct^2 = 0$. That is, every line through P, with the single exception of the line

$$\tau_{\mathscr{P}} \equiv r^2x - 2try + t^2z = 0$$

meets Γ in exactly one other point, while $\tau_{\mathscr{P}}$ meets Γ only at P.

DEFINITION 2.6.3 *Tangent* to a conic Γ at a point P of Γ: the unique line through P which meets Γ only at P.

From the discussion above it follows that:

THEOREM 2.6.2 (i) $P \in \mathscr{C}P \Leftrightarrow P \in \Gamma$, (ii) $P \in \mathscr{C}P \Leftrightarrow \mathscr{C}P$ *is the tangent to* Γ *at* P.

(Any phrase such as '$\tau_{\mathscr{P}}$ meets Γ twice at P' is quite unacceptable. If $\mathscr{K}$ is a finite field of order n, then Γ consists of $n+1$ points, namely, the n points $(t^2, t, 1)$ together with $(1, 0, 0)$. Through a point P of Γ there are $n+1$ lines; each of n of them contains exactly one other point of Γ, while the last meets Γ only at P.)

Consider next the relation of the polar line of $P = (a, b, c)$ to Γ. $\mathscr{C}P$ is $cx - 2by + az = 0$ and meets Γ, if at all, in the points (t^2, tr, r^2) which satisfy the condition

$$ct^2 - 2btr + ar^2 = 0.$$

That is, the polar of P meets the conic if and only if in the field $\mathscr{K}$ this equation has roots, namely if $b^2 - ac$ is a square in $\mathscr{K}$. Assume then that $\mathscr{C}P \cap \Gamma = \{Q, R\}$. $Q \in \mathscr{C}P \Rightarrow P \in \mathscr{C}Q$; that is, P lies on the tangents to Γ at Q and R. Using this property we can split the points of the plane into three disjoint subsets, which, borrowing the terminology from that used for an ellipse in the real plane, we may define as follows:

DEFINITION 2.6.4 For a conic Γ: *exterior point*, P_e: P_e is the intersection of two tangents to Γ. Point P_0 on Γ, or *absolute point* in the polarity $\mathscr{C}$: P_0 lies on a single tangent to Γ. *Interior point* P_i: P_i lies on no tangent to Γ.

THEOREM 2.6.3 *If* Γ *is* $y^2 - xz = 0$, *then*

$$\{P_e\} = \{(x_e, y_e, z_e) \colon y_e^2 - x_e z_e \text{ is a non-zero square in } \mathscr{K}\},$$

$$\{P_0\} = \{(x_0, y_0, z_0) \colon y_0^2 - x_0 z_0 = 0\},$$

$$\{P_i\} = \{(x_i, y_i, z_i) \colon y_i^2 - x_i z_i \text{ is a not-square in } \mathscr{K}\}.$$

Let us now consider the general non-singular quadratic form in (x_1, x_2, x_3); $\mathbf{x}^T\mathbf{C}\mathbf{x} = \Sigma c_{ij} x_i x_j$, $c_{ij} = c_{ji}$, $\det \mathbf{C} \neq 0$. In relation to $\mathbf{C}$ we frame the following alternative definition of a conic Γ:

DEFINITION 2.6.5 *Non-singular conic*† in $\Pi(\mathscr{K})$, char $\mathscr{K} \neq 2$: the set of points $\{\mathbf{x} \colon \mathbf{x}^T\mathbf{C}\mathbf{x} = 0\}$, where $\mathbf{C}$ is a non-singular symmetric matrix over $\mathscr{K}$.

In the following discussion we take account only of non-singular conics, excluding from consideration the conic (line-pair, single line or single point) determined by a singular matrix $\mathbf{C}$.

EXERCISE 2.6.1 Show that any line meets a non-singular conic at either two, one or no points.

DEFINITION 2.6.6 *Polarity* in a projective plane Π: a 1–1 correspondence $P \to \mathscr{M}P$ between the points of Π and the lines of Π such that $Q \in \mathscr{M}P \Rightarrow P \in \mathscr{M}Q$. $\mathscr{M}P$ is the *polar* of P and P is the *pole* of $\mathscr{M}P$.

We write also $\mathscr{M}l$ for the pole of the line l.

THEOREM 2.6.4 *In a polarity every set of collinear points corresponds to a set of concurrent lines.*

Let $\{Q_i\} \subset l$ and let $l = \mathscr{M}P$, then $Q_i \in l \Rightarrow P \in \mathscr{M}Q_i$; that is, the polar lines of points in the collinear set $\{Q_i\}$ are lines of a set concurrent in P.

EXERCISE 2.6.2 Prove that in a polarity every set of concurrent lines corresponds to a set of collinear points.

† Since the set may be empty, this definition is wider than Definition 2.6.1 but only slightly wider: see Exercise 2.6.3.

The quadratic form $\mathbf{x}^T\mathbf{C}\mathbf{x}$ defines a 1–1 correspondence $\mathbf{p} \to \mathscr{C}\mathbf{p}$, where $\mathscr{C}\mathbf{p}$ is $\mathbf{p}^T\mathbf{C}\mathbf{x} = 0$, between the points and the lines of $\Pi(\mathscr{K})$. Since $\mathbf{C}$ is symmetric

$$\mathbf{q} \in \mathscr{C}\mathbf{p} \Rightarrow \mathbf{p}^T\mathbf{C}\mathbf{q} = 0 \Rightarrow \mathbf{q}^T\mathbf{C}\mathbf{p} = 0 \Rightarrow \mathbf{p} \in \mathscr{C}\mathbf{q}$$

so that the correspondence is a polarity. Let $\mathbf{p}$ be any point not on the conic Γ, $\mathbf{x}^T\mathbf{C}\mathbf{x} = 0$, and let $\mathbf{q}$ be any point of $\mathscr{C}\mathbf{p}$ which is not on Γ (cf. Exercise 2.6.1). Take $\mathbf{r} = \mathscr{C}\mathbf{p} \cap \mathscr{C}\mathbf{q}$, then $\mathbf{p}$, $\mathbf{q}$, $\mathbf{r}$ are non-collinear and form a triangle each of whose sides is the polar of the opposite vertex. Such a triangle is called *self-polar*. Now change the coordinate-system so that $\mathbf{p}$, $\mathbf{q}$, $\mathbf{r}$ becomes the triangle of reference, and assume that the matrix $\mathbf{C}$ is transformed to $\mathbf{C}'$. Since in this system the polar of $(1, 0, 0)$ for $\mathbf{x}^T\mathbf{C}'\mathbf{x}$ is $x_1 = 0$, we have $c'_{12} = c'_{13} = 0$; likewise $c'_{23} = 0$. Thus:

THEOREM 2.6.5 *For any non-singular conic Γ there exist self-polar triangles. If the reference triangle is self-polar for Γ, then the equation of Γ reduces to $ax^2 + by^2 + cz^2 = 0$ with none of a, b, c zero.*

We prove next:

THEOREM 2.6.6 *If there is one point on the non-singular conic $\mathbf{x}^T\mathbf{C}\mathbf{x} = 0$ (that is, if the associated polarity has one absolute point) then the points of Γ (absolute points) are in 1–1 correspondence with the points of a line.*

Since there is one point for which $\mathbf{x}^T\mathbf{C}\mathbf{x} = 0$, we may select it to be $(1, 0, 0)$ and then $c_{11} = 0$. The line $lx_2 = mx_3$ meets Γ in $(1, 0, 0)$ and in the point

$$(c_{22}m^2 + 2c_{23}lm + c_{33}l^2, \quad -2m(c_{12}m + c_{13}l), \quad -2l(c_{12}m + c_{13}l)).$$

Thus every line through $(1, 0, 0)$ except one meets Γ in exactly one other point, the exceptional line being $c_{12}x_2 + c_{13}x_3 = 0$, which meets Γ only in $(1, 0, 0)$. Every point of the plane lies on some line through $(1, 0, 0)$ so that all points of Γ have been accounted for.

There may of course be no point for which the quadratic form $\mathbf{x}^T\mathbf{C}\mathbf{x}$ vanishes in the field $\mathscr{K}$. For example, $x_1^2 + x_2^2 + x_3^2$ has no zero in the field of real numbers. However, it can be proved that over any finite field (of odd characteristic), say a field of order n, every form $\mathbf{x}^T\mathbf{C}\mathbf{x}$ has a zero and therefore has $n+1$ zeroes. We shall prove this later for the case $n = 9$. A quadratic form that has no zero determines a polarity, but there are no absolute points, and therefore no separation of the points of the plane into 'interior' and 'exterior'. On the other hand, in the plane over a field which is algebraically closed, there are absolute points associated with every quadratic form, but there is no separation of the non-absolute points into 'exterior' and 'interior'.

EXERCISE 2.6.3 If a non-singular quadratic form $\mathbf{x}^T\mathbf{C}\mathbf{x}$ in (x, y, z) has a zero in the field $\mathscr{K}$, char $\mathscr{K} \neq 2$, then coordinate-systems can be found in which the equation $\mathbf{x}^T\mathbf{C}\mathbf{x} = 0$ can be reduced to $y^2 - xz = 0$.

As a consequence of this result, if $\mathbf{x}^T\mathbf{C}\mathbf{x}$ has at least one zero in $\mathscr{K}$, we can find a constant k and a matrix $\mathbf{M}$ such that

$$\mathbf{x}^T(k\mathbf{C})\,\mathbf{x} \equiv \mathbf{x}^T\mathbf{M}^T\mathbf{R}\mathbf{M}\mathbf{x},$$

where
$$\mathbf{x}^T\mathbf{R}\mathbf{x} \equiv y^2 - xz;$$

that is,
$$k\mathbf{C} = \mathbf{M}^T\mathbf{R}\mathbf{M}$$

and
$$k^3 \det \mathbf{C} = -\tfrac{1}{4}(\det \mathbf{M})^2.$$

It follows that $\det \mathbf{C}$ is a square in $\mathscr{K}$ if and only if $-k$ is a square, and therefore that the sets

$$\{\mathbf{x} : y^2 - xz \text{ is a square}\} \quad \text{and} \quad \{\mathbf{x} : (-\det \mathbf{C}\,.\,\mathbf{x}^T\mathbf{C}\mathbf{x}) \text{ is a square}\}$$

are identical.

THEOREM 2.6.7 *If $\mathbf{x}^T\mathbf{C}\mathbf{x}$ is a quadratic form with at least one zero in $\mathscr{K}$, the sets of points for which $-\det \mathbf{C}\,.\,\mathbf{x}^T\mathbf{C}\mathbf{x}$ is a non-zero square, zero and a not-square are respectively the points exterior to, on and interior to the conic $\mathbf{x}^T\mathbf{C}\mathbf{x} = 0$.*

EXERCISE 2.6.4 By selecting coordinates so that one of two non-singular conics has the equation $y^2 - xz = 0$, prove that two non-singular conics may have 4, 3, 2, 1 or 0 common points.

EXERCISE 2.6.5 Prove that the polarity with regard to a conic Γ establishes the following correspondences:

(i) Point of $\Gamma \leftrightarrow$ tangent line to Γ.

(ii) Exterior point of $\Gamma \leftrightarrow$ chord of Γ.

(iii) Interior point of $\Gamma \leftrightarrow$ non-secant of Γ, where a chord contains two points of Γ and a non-secant contains none.

EXERCISE 2.6.6 Prove that in relation to a non-singular conic $y^2 = xz$ over a finite field of odd order n there are $\frac{1}{2}n(n+1)$ exterior points and $\frac{1}{2}n(n-1)$ interior points.

EXERCISE 2.6.7 Prove that the set of interior points determined by a non-singular conic $y^2 = xz$ in the plane over GF(5) forms a Desargues figure.

2.7 Projective correlations of $\Pi(\mathscr{K})$

DEFINITION 2.7.1 A *correlation* $\mathscr{M}$ is a 1–1 correspondence $P \to \mathscr{M}P$ between the set of points of a plane and the set of lines of the plane such that every collinear set of points corresponds to a concurrent set of lines and every concurrent set of lines corresponds to a collinear set of points.

A special type of correlation is the polarity determined by a quadratic form.

A plane is self-dual if and only if it possesses at least one correlation. Since $\Pi(\mathscr{K})$ has polarities, we have :

THEOREM 2.7.1 *The plane* $\Pi(\mathscr{K})$ *is self-dual.*

The following theorem is an immediate consequence of the definition of collineations and correlations:

THEOREM 2.7.2 *If, in a projective plane,* $\mathscr{C}$ *is a collineation and* $\mathscr{L}$ *and* $\mathscr{M}$ *are correlations, then* $\mathscr{L}^{-1}$, $\mathscr{L}\mathscr{C}$ *and* $\mathscr{C}\mathscr{L}$ *are correlations and* $\mathscr{L}\mathscr{M}$ *is a collineation.*

It follows that if $\mathscr{L}$, $\mathscr{M}$ are any two correlations then there exist collineations $\mathscr{C}$, $\mathscr{D}$ such that $\mathscr{M} = \mathscr{L}\mathscr{C} = \mathscr{D}\mathscr{L}$, namely $\mathscr{C} = \mathscr{L}^{-1}\mathscr{M}$, $\mathscr{D} = \mathscr{M}\mathscr{L}^{-1}$.

THEOREM 2.7.3 *If $\{\mathscr{C}\}$ is the set of collineations of a projective plane, and $\mathscr{L}$ is any given correlation, then $\{\mathscr{C}\mathscr{L}\}$ is the set of correlations in the plane, and $\{\mathscr{C}\mathscr{L}\}$ and $\{\mathscr{C}\}$ are in* 1–1 *correspondence.*

The correlations cannot, of course, form a group by themselves, but correlations and collineations together form a group under geometrical composition.

From the algebraic point of view the simplest correlation in the plane $\Pi(\mathscr{K})$ is $\mathscr{U}$, namely

$$\mathscr{U}: \mathbf{a} \rightarrow \mathbf{a}^T\mathbf{x} = 0.$$

If $\mathscr{U}$ is compounded with a projectivity

$$\mathscr{A}: \mathbf{a} \rightarrow \mathbf{Aa}$$

we obtain the correlation

$$\mathscr{U}\mathscr{A}: \mathbf{a} \rightarrow (\mathbf{Aa})^T\mathbf{x} = 0 \equiv \mathbf{a}^T\mathbf{A}^T\mathbf{x} = 0.$$

DEFINITION 2.7.2 The set of *projective correlations* in $\Pi(\mathscr{K})$ is $\{\mathscr{M}: \mathbf{a} \rightarrow \mathbf{a}^T\mathbf{M}\mathbf{x} = 0\ (\mathbf{M} \text{ non-singular})\}$.

EXERCISE 2.7.1 What is the equation of the line corresponding to a point $\mathbf{a}$ in the correlation $\mathscr{A}\mathscr{U}$?

Our next objective is to devise a geometrical construction for a projective correlation in $\Pi(\mathscr{K})$; since we know already that every projectivity is the resultant of a set of geometric operations (homologies and elations), we have only to show that one particular correlation can be constructed geometrically. The correlation $\mathscr{U}$ is not the simplest from this point of view, and (again for our present purpose) has the disadvantage that it is the polarity determined by the quadratic form $\mathbf{x}^T\mathbf{x}$. Let us consider instead the correlation

$$\mathscr{T}: (u, v, w) \rightarrow [w, u, v]\,\mathbf{x} = 0.$$

The correlative line $\mathscr{T}P$ of a point P can be constructed in the following way (Diagram 2.7):

$$X, Y, Z, K = (1, 0, 0), (0, 1, 0), (0, 0, 1), (1, 1, 1),$$

$$D = YZ \cap XK = (0, 1, 1), \quad E = (1, 0, 1), \quad F = (1, 1, 0).$$

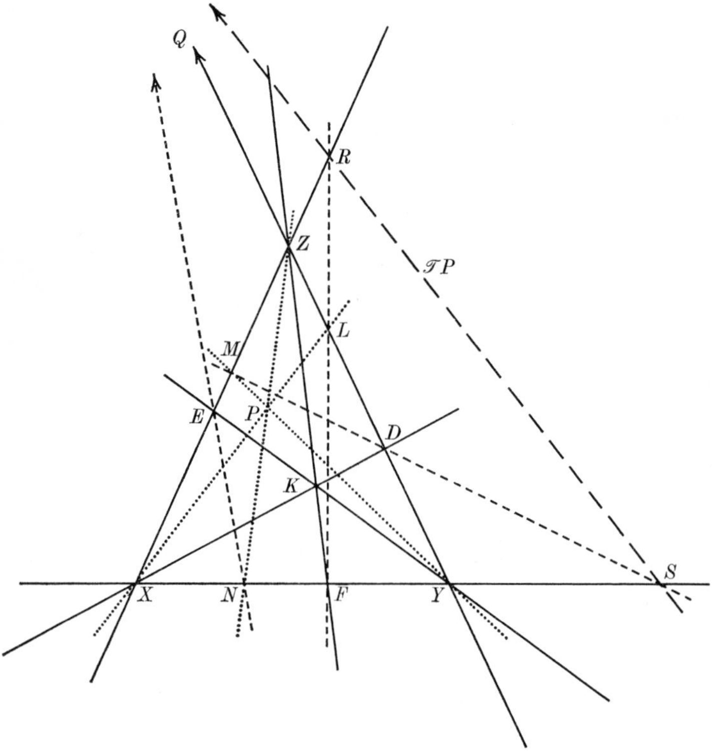

Diagram 2.7

(The case in which D, E, F are collinear, the characteristic of $\mathscr{K}$ being two, is not excluded.)

$P = (u, v, w)$ is an arbitrary point,

$L = XP \cap YZ = (0, v, w), \quad M = (u, 0, w), \quad N = (u, v, 0),$

$Q = EN \cap YZ = (u, v, 0) - u(1, 0, 1) = (0, v, -u),$

$R = FL \cap ZX = (-v, 0, w), \quad S = DM \cap XY = (u, -w, 0).$

QR is $wx + uy + vz = 0$, so that $QR = \mathscr{T}P$. Also $S \in QR$.

The proof above is algebraic; a geometrical proof could be devised, but it would require in effect the construction of the correlative lines of three collinear points, and the proof of the concurrence of these three lines; it is therefore necessarily very involved. In the course of the algebraic proof we have to make use of the commutative law of multiplication. The geometric proof of the theorem must therefore depend on the Pappus axiom.

EXERCISE 2.7.2 Determine the coordinates of the points

(i) $\mathscr{T}(QR) = \mathscr{T}Q \cap \mathscr{T}R$,

(ii) $\mathscr{T}(px+qy+rz = 0)$.

In the same way as in Theorem 2.4.6 we may prove that a projective correlation of $\Pi(\mathscr{K})$ is uniquely determined when a regular ordered quadrangle is assigned to be the correlative of a regular ordered quadrilateral. For the correlations $\mathscr{T}$ and $\mathscr{U}$ the assigned elements could be

$$\mathscr{T}\colon (X, Y, Z, K) \to (XZ,\ YX,\ ZY,\ k),$$

$$\mathscr{U}\colon (X, Y, Z, K) \to (YZ, ZX, XY, k),$$

where $\quad K = (1, 1, 1) \quad$ and $\quad k$ is $\quad x+y+z = 0.$

DEFINITION 2.7.3 *Absolute point* for a correlation: a point which lies on its correlative line.

For a given projective correlation, $\mathscr{M}\colon \mathbf{a} \to \mathbf{a}^T M\mathbf{x} = 0$, the absolute points form the set $\{\mathbf{x}\colon \mathbf{x}^T\mathbf{M}\mathbf{x} = 0\}$. For the correlations $\mathscr{T}$ and $\mathscr{U}$ the sets are

$$\mathscr{T}_*\colon \{\mathbf{x}\colon yz+zx+xy = 0\},$$

$$\mathscr{U}_*\colon \{\mathbf{x}\colon x^2+y^2+z^2 = 0\}.$$

For $\mathscr{T}$ the set is never empty, whatever the field, while for $\mathscr{U}$, over the real field at least, the set is empty.

Now let us look a little more closely at the structure of the relation satisfied by the absolute points of $\mathscr{M}$:

$$\mathbf{x}^T\mathbf{M}\mathbf{x} = m_{11}x_1^2+m_{22}x_2^2+m_{33}x_3^2+(m_{23}+m_{32})\,x_2x_3$$
$$+(m_{31}+m_{13})\,x_1x_2+ +(m_{12}+m_{21})\,x_1x_2.$$

Write $\quad c_{ij} = c_{ji} = \tfrac{1}{2}(m_{ij}+m_{ji}),$

$$s_{ij} = -s_{ji} = \tfrac{1}{2}(m_{ij}-m_{ji}),$$

say $\mathbf{M} = \mathbf{C} + \mathbf{S}$ where $\mathbf{C}^T = \mathbf{C}$, $\mathbf{S}^T = -\mathbf{S}$.

Then
$$\mathbf{x}^T\mathbf{M}\mathbf{x} = \mathbf{x}^T\mathbf{C}\mathbf{x} + \mathbf{x}^T\mathbf{S}\mathbf{x} = \mathbf{x}^T\mathbf{C}\mathbf{x}.$$

THEOREM 2.7.4 *If* $\mathbf{M} = \mathbf{C} + \mathbf{S}$, *where* $\mathbf{C} = \frac{1}{2}(\mathbf{M} + \mathbf{M}^T)$, $\mathbf{S} = \frac{1}{2}(\mathbf{M} - \mathbf{M}^T)$, *then the set of absolute points of the correlation* $\mathbf{a} \to \mathbf{a}^T\mathbf{M}\mathbf{x} = 0$ *is* $\{\mathbf{x}: \mathbf{x}^T\mathbf{C}\mathbf{x} = 0\}$, *and is independent of the values assigned to the three distinct non-zero elements of* $\mathbf{S}$.

If $\mathbf{S} = \mathbf{0}$, then the correlation is a *polarity*, $\mathscr{C}$, *wth regard to a conic.* There are two important differences between a polarity $\mathscr{C}$ and a general projective correlation $\mathscr{M}$:

(i) If a point P lies on the conic $\mathbf{x}^T\mathbf{M}\mathbf{x} = 0$, then $P \in \mathscr{M}P$, but $\mathscr{M}P$ is not the tangent to the conic at P.

(ii) The symmetrical relation characteristic of polarities, $P' \in \mathscr{C}P \Rightarrow P \in \mathscr{C}P'$, no longer holds. For example in the correlation $\mathscr{T}$, with matrix

$$\mathbf{T} = \begin{bmatrix} 0 & 0 & 1 \\ 1 & 0 & 0 \\ 0 & 1 & 0 \end{bmatrix} = \frac{1}{2}\begin{bmatrix} 0 & 1 & 1 \\ 1 & 0 & 1 \\ 1 & 1 & 0 \end{bmatrix} + \frac{1}{2}\begin{bmatrix} 0 & -1 & 1 \\ 1 & 0 & -1 \\ -1 & 1 & 0 \end{bmatrix},$$

the correlative of X is $z = 0$, which contains the two points X and Y of the conic $yz + zx + xy = 0$. If $P = (a, b, c)$, $P' = (a', b', c')$ then
$$P' \in \mathscr{T}P \Leftrightarrow ca' + ab' + bc' = 0,$$
but
$$P \in \mathscr{T}P' \Leftrightarrow ba' + cb' + ac' = 0.$$

In more general terms, under the correlation $\mathbf{p} \to \mathbf{p}^T\mathbf{M}\mathbf{x} = 0$, the point of which $\mathbf{u}^T\mathbf{x} = 0$ is the correlative line is $(\mathbf{M}^T)^{-1}\mathbf{u}$, while the correlative point of the line $\mathbf{u}^T\mathbf{x} = 0$ is $\mathbf{M}^{-1}\mathbf{u}$, that is, $\mathbf{M}^{-1}\mathbf{u}$ is the common point of the correlative lines of the points of $\mathbf{u}^T\mathbf{x} = 0$.

CHAPTER 3

GALOIS PLANES OF ORDERS 3 AND 9

3.1 The construction of the plane Δ of order 3

As a preliminary to the construction and investigation of the Galois planes of orders 3 and 9 we consider again the configuration constructed in the first few steps of completing a quadrangle. This configuration plays an important part not only in these and other field-planes but also in one of the miniquaternion planes. In any projective plane: $ABCD$ is a given regular quadrangle, EFG is the diagonal triangle of $ABCD$. The seven points constructed so far form a Fano plane if and only if E, F, G are collinear. We assume that they are not collinear, and construct the points where the sides of the quadrangle meet the sides of the triangle:

$$L = BC \cap FG, \quad M = CA \cap GE, \quad N = AB \cap EF,$$

$$L' = AD \cap FG, \quad M' = BD \cap GE, \quad N' = CD \cap EF.$$

So far we have 13 points and 9 lines with four points on each line and four or fewer lines through each point. Because $ABCD$ was defined to be a regular quadrangle and E, F, G were prescribed to be not collinear, no two of these points and no two of these lines can coincide. Thus, if any projective plane exists which contains $ABCD$ and is not a Fano plane, it must contain these 13 points and 9 lines; in particular the 13 points must be the complete set of points in any plane of order 3 that contains $ABCD$. If we assume that such a plane exists, then it has to be completed by prescribing four more sets of four collinear points. LM must contain one of A, D, E or L', and clearly cannot contain A (because $L \neq C$), or L' (because $G \neq M$), or E. Therefore, in a plane of order 3, LM contains D. By similar reasoning we can prove that the remaining four collinear sets have to be $DLMN$, $ALM'N'$, $BMN'L'$, $CNL'M'$. If we prescribe these collinearities we have a system of 13 points collinear by sets of 4 and 13 lines

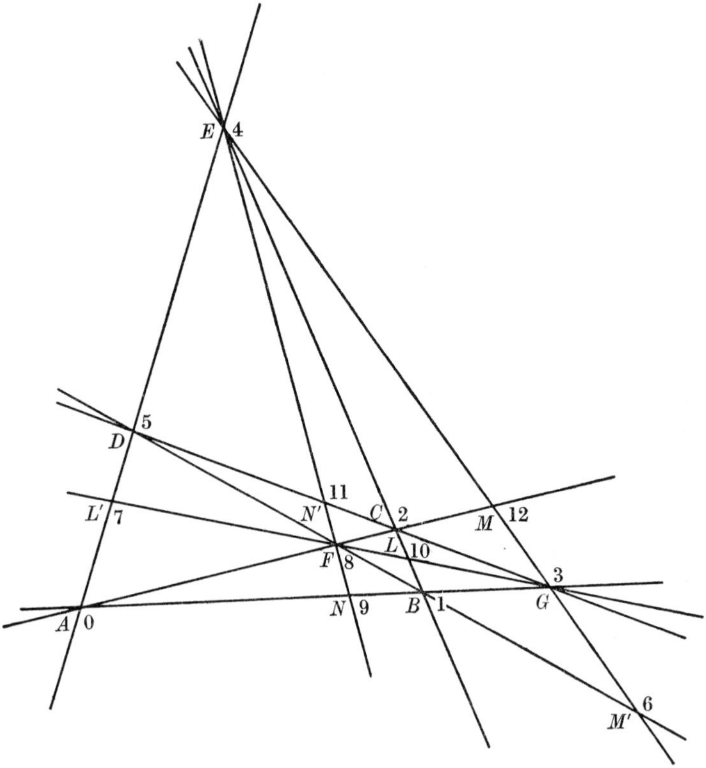

Diagram 3.1

concurrent by sets of 4, that is, we have constructed from the given regular quadrangle a plane of order 3. At every step the construction was uniquely defined, so that:

THEOREM 3.1.1 *There is a unique plane of order* 3, *that is, any two planes of order* 3 *are isomorphic.*†

NOTATION 3.1.1 Δ: the plane of order 3.

† It has been proved that the planes of orders 4, 5, 7 and 8 are unique (the first three by hand and the last by computer), and that there is no plane of order 6. There are at least four non-isomorphic planes of order 9 and these form the main subject of this book. It is not known whether there is any other plane of order 9, nor whether there is any plane of order 10. The only known finite planes are of prime-power order. (See pp. 172–3.)

The pattern of incidences in Diagram 3.1 is most easily appreciated if we arrange the sets of collinear points in columns in the following way:

Table 3.1.1

B	C	A	A	B	C	F	G	E	A	B	C	D
C	A	B	D	D	D	G	E	F	L	M	N	L
E	F	G	E	F	G	L	M	N	M'	N'	L'	M
L	M	N	L'	M'	N'	L'	M'	N'	N'	L'	M'	N

As for the Fano plane, an even clearer picture is obtained if we name the points K_r, $r = 0, \ldots, 12$, in such a way that we can construct a cyclic table. Thus take:

	A	B	C	D	E	F	G	L	M	N	L'	M'	N'
$K_r: r =$	0	1	2	5	4	8	3	10	12	9	7	6	11

and then rearrange the rows and columns of Table 3.1.1 and name the lines k_s appropriately so as to produce the following table:

Table 3.1.2 *Cyclic table for* Δ

$K_r: r =$	0	1	2	3	4	5	6	7	8	9	10	11	12
	1	2	3	4	5	6	7	8	9	10	11	12	0
	3	4	5	6	7	8	9	10	11	12	0	1	2
	9	10	11	12	0	1	2	3	4	5	6	7	8
$k_s: s =$	0	1	2	3	4	5	6	7	8	9	10	11	12

$k_s = \{K_s, K_{s+1}, K_{s+3}, K_{s+9}\}$, mod 13

$K_r = k_r \cap k_{r-1} \cap k_{r-3} \cap k_{r-9}$, mod 13

EXERCISE 3.1.1 Check that the incidence scheme in Table 3.1.2 satisfies the conditions for a projective plane.

From the cyclic table it is easy to find some collineations of Δ. For example, $\mathscr{A}: K_r \to K_{r+1}$ (mod 13) permutes the columns of the table, as does $\mathscr{B}: K_r \to K_{3r}$ (mod 13).

EXERCISE 3.1.2 Show that $\mathscr{B}\mathscr{A} = \mathscr{A}^3\mathscr{B}$.

Let us assume now that the configuration of Diagram 3.1 has been constructed in the plane $\Pi(\mathscr{K})$ over a commutative field $\mathscr{K}$, and that homogeneous coordinates have been assigned so that

$$A = (1,0,0), \quad B = (0,1,0), \quad C = (0,0,1), \quad D = (1,1,1).$$

The coordinates of the remaining nine points are then

$$\begin{array}{lll} E:(0,1,1) & L:(0,-1,1) & L':(2,1,1) \\ F:(1,0,1) & M:(1,0,-1) & M':(1,2,1) \\ G:(1,1,0) & N:(-1,1,0) & N':(1,1,2). \end{array}$$

In $\Pi(\mathscr{K})$ the four sets of points LMN, $LM'N'$, $MN'L'$, $NL'M'$ are collinear—one set lies on each of the lines $x \pm y \pm z = 0$. The points D, L, M are collinear if and only if $(1,1,1)$ lies on $x+y+z = 0$; that is, the field $\mathscr{K}$ is of characteristic 3. Likewise, AL is $y+z = 0$ and contains M' if $2+1 = 0$, and $BM \supset N'$ and $CN \supset L'$ impose the same condition, so that:

THEOREM 3.1.2 *The completion of a regular quadrangle in* $\Pi(\mathscr{K})$ *is a subplane of order* 3 *if and only if* $\mathscr{K}$ *is of characteristic* 3.

EXERCISE 3.1.3 Show that the completion of a regular quadrangle in $\Pi(\mathscr{K})$ is a Fano subplane if and only if $\mathscr{K}$ is of characteristic 2.

3.2 The collineation group of Δ

The completion of any regular quadrangle in Δ is, by Theorem 3.1.2, the whole plane Δ. It follows that any collineation of Δ is uniquely determined by the points to which the vertices of a regular quadrangle are mapped. By Theorem 2.4.6 there is a projectivity which maps any ordered regular quadrangle onto any other, so that

THEOREM 3.2.1 *Every collineation of* Δ *is a projectivity.*

From this it follows, by Theorem 2.4.4, that

THEOREM 3.2.2 *Every collineation of* Δ *can be represented by a relation* $\mathbf{x} \to \mathbf{Mx}$, *where* $\mathbf{M}$ *is a non-singular* 3×3 *matrix over* $\mathscr{D}$.

EXERCISE 3.2.1 Prove that the number of different unordered regular quadrangles in Δ is 234. What is the order of the group of collineations?

Valuable insight into the structure of a Galois plane is obtained by representing the points of the plane as a sequence $\mathbf{p}$, $\mathbf{Hp}$, $\mathbf{H}^2\mathbf{p}, \ldots$ where $\mathbf{p}$ is the coordinate-vector of some fixed point, and $\mathbf{H}$ is a matrix representing a projectivity.

DEFINITION 3.2.1 *Singer cycle*: a sequence of points $\mathbf{p}$, $\mathbf{Hp}$, $\mathbf{H}^2\mathbf{p}, \ldots$ which includes all points of the plane. $\mathbf{H}$ is a *Singer matrix*.†

If the sequence $\mathbf{p}, \mathbf{Hp}, \mathbf{H}^2\mathbf{p}, \ldots$ is to contain all thirteen points of the plane Δ, then clearly $\mathbf{H}$ can have no eigenvalue, for, if e is an eigenvalue, and $\mathbf{Hu} = e\mathbf{u}$, then either $\mathbf{p} = k\mathbf{u}$ for some k and the sequence contains only the point $\mathbf{u}$, or else the sequence does not include the point $\mathbf{u}$. We are concerned therefore with matrices $\mathbf{H}$ such that the characteristic polynomial $\chi(\mathbf{H})$ is irreducible.

The projectivities represented by the matrices $\mathbf{M}$ and $-\mathbf{M}$ are identical; $\det(-\mathbf{M}) = -\det \mathbf{M} = \pm 1$ (since $\mathbf{M}$ is non-singular and all the coefficients of $\mathbf{M}$ belong to $\mathscr{D}$) so we may confine our attention to matrices $\mathbf{H}$ such that $\det \mathbf{H} = 1\,(= \det \mathbf{H}^r)$. It follows, since the constant term in $\chi(\mathbf{H})$ is $-\det \mathbf{H}$, that the complete set of relevant characteristic polynomials is the set of nine:

$$\chi(\mathbf{H}) \equiv x^3 + lx^2 + mx - 1 \quad (l, m \in \mathscr{D}).$$

The reducible cubic polynomials of this form over $\mathscr{D}$ are

$$(x-1)^3 = x^3 - 1,$$

$$(x-1)(x+1)^2 = x^3 + x^2 - x - 1,$$

$$(x+1)(x^2+x-1) = x^3 - x^2 - 1,$$

$$(x-1)(x^2+1) = x^3 - x^2 + x - 1,$$

$$(x+1)(x^2-x-1) = x^3 + x - 1.$$

† J. Singer (1938). Not every finite plane is such that its points can be arranged as a Singer cycle, but it can be proved that in every Galois plane such an arrangement is possible.

There are therefore four irreducible polynomials and we find that these form a single cycle under the 'squaring process' described in Theorem 2.5.4, namely

$$x^3+lx^2+mx-1 \to x^3+(2m-l^2)\,x^2+(m^2+2l)\,x-1.$$

Thus, if $\chi(\mathbf{H})$ is irreducible, then $\chi(\mathbf{H})$, $\chi(\mathbf{H}^2)$, $\chi(\mathbf{H}^4)$, $\chi(\mathbf{H}^8)$ are the four irreducible polynomials of the form x^3+lx^2+mx-1.

Each of these four polynomials is therefore as suitable a candidate as the others. Consider a matrix $\mathbf{G}$ of which the characteristic polynomial is

$$\chi(\mathbf{G}) \equiv x^3-x-1.$$

Without loss of generality, by Theorem 2.5.3, we may take $\mathbf{G}$ to be

$$\mathbf{G} = \begin{bmatrix} 0 & 0 & 1 \\ 1 & 0 & 1 \\ 0 & 1 & 0 \end{bmatrix}.$$

By the Cayley–Hamilton Theorem (Lemma 1 of Theorem 2.5.3)

$$\mathbf{G}^3-\mathbf{G}-\mathbf{1} = \mathbf{0}.$$

So
$$\mathbf{G}^9 = (\mathbf{G}+\mathbf{1})^3 = \mathbf{G}^3+\mathbf{1} = \mathbf{G}+\mathbf{1}+\mathbf{1} = \mathbf{G}-\mathbf{1},$$
$$\mathbf{G}^{12} = \mathbf{G}^3\mathbf{G}^9 = (\mathbf{G}+\mathbf{1})(\mathbf{G}-\mathbf{1}) = \mathbf{G}^2-\mathbf{1},$$
$$\mathbf{G}^{13} = \mathbf{G}\mathbf{G}^{12} = \mathbf{G}(\mathbf{G}^2-\mathbf{1}) = \mathbf{G}^3-\mathbf{G} = \mathbf{1}.$$

Since, for any matrix $\mathbf{K}$ over GF(3), $\chi(\mathbf{K}) = \chi(\mathbf{K}^3)$ (Theorem 2.5.4) and $\mathbf{G}^{13} = \mathbf{1}$, we have, writing $\chi_r = \chi(\mathbf{G}^r)$,

$$\chi_1 = \chi_3 = \chi_9 = x^3-x-1,$$
$$\chi_2 = \chi_6 = \chi_5 = x^3+x^2+x-1,$$
$$\chi_4 = \chi_{12} = \chi_{10} = x^3+x^2-1,$$
$$\chi_8 = \chi_{11} = \chi_7 = x^3-x^2-x-1.$$

$\mathbf{G}$ is therefore a Singer matrix; it determines the cycle of points shown in Table 3.2.1.

Table 3.2.1 *Vectors* $\acute{\mathbf{g}}_r = \mathbf{G}^r(1,0,0)$ *and matrices* $\mathbf{G}^r$

$\acute{\mathbf{g}}_r$, $r =$	0	1	2	3	4	5	6	7	8	9	10	11	12	(0	1)
$\mathbf{G}^r \begin{bmatrix}1\\0\\0\end{bmatrix}$	1	0	0	1	0	1	1	1	−1	−1	0	1	−1	1	0
	0	1	0	1	1	1	−1	−1	0	1	−1	1	0	0	1
	0	0	1	0	1	1	1	−1	−1	0	1	−1	1	0	0

$\mathbf{G}^r = [\acute{\mathbf{g}}_r, \acute{\mathbf{g}}_{r+1}, \acute{\mathbf{g}}_{r+2}]$, mod 13.

EXERCISE 3.2.2 Explain why $\mathbf{G}^r = [\mathbf{g}_r, \mathbf{g}_{r+1}, \mathbf{g}_{r+2}]$.

EXERCISE 3.2.3 Read by rows, Table 3.2.1 is such that the second and third rows are cyclic permutations of the first. Explain this property in terms of the structure of the matrix $\mathbf{G}$.

EXERCISE 3.2.4 Let $K_r = \mathbf{G}^r(1, 0, 0)$. Verify that K_0, K_1, K_3, K_9 are collinear and prove that in consequence $K_r, K_{r+1}, K_{r+3}, K_{r+9}$ are collinear, and the collinear sets form the cyclic Table 3.1.2.

EXERCISE 3.2.5 Write $k_r = \{K_r, K_{r+1}, K_{r+3}, K_{r+9}\}$, mod 13. Prove that:

(i) $K_s = k_s \cap k_{s+12} \cap k_{s+10} \cap k_{s+4}$, mod 13.

(ii) The coordinate-vector of the line k_r is given by

$$\mathbf{k}_r^T = [e_r, e_{r-1}, e_{r-2}]$$

in the following table:

$r =$	12	11	10	9	8	7	6	5	4	3	2	1	0	(12 | 11)
$e_r =$	0	1	0	1	1	1	−1	−1	0	1	−1	1	0	0	1

(for example $\mathbf{k}_7^T = [1, -1, -1]$). The sequence of numbers $e_0, \ldots, e_{12}$ forms the row of second coordinates in Table 3.2.1. Why?

EXERCISE 3.2.6

$$\mathbf{T} \overset{\text{def}}{=} \begin{bmatrix} 1 & 0 & 0 \\ 0 & 0 & 1 \\ 0 & 1 & 0 \end{bmatrix}.$$

Prove that $\mathbf{TG}^r$ is symmetric and explain this property in terms of the result in Exercise 3.2.3.

EXERCISE 3.2.7 Prove that in Δ: (i) a non-trivial homology is completely determined by its axis and centre, (ii) a given axis and centre determine two non-trivial elations, one the inverse of the other.

EXERCISE 3.2.8 Prove that the projectivity determined by $\mathbf{G}$ is the resultant of the sequence of central collineations

$$\mathscr{G} : \mathscr{C}_{1,1;7 \to 12}\mathscr{C}_{9,2}\mathscr{C}_{10,4},$$

where $\mathscr{C}_{i,j} \neq 1$ is a central collineation with centre K_i and axis k_j, and $\mathscr{C}_{i,j;p\to q}$ is the elation in which $K_p \to K_q$. Prove that the following construction determines $\mathscr{G}P$ from P:

$$Q = PK_{10} \cap \langle K_6, k_4 \cap PK_{11} \rangle,$$

$$R = QK_9 \cap \langle K_7, k_2 \cap QK_6 \rangle,$$

$$\mathscr{G}P = RK_1 \cap \langle K_{12}, k_1 \cap RK_7 \rangle.$$

EXERCISE 3.2.9 Prove that the collineation $K_r \to K_{3r}$ is the projectivity represented by the matrix

$$\begin{bmatrix} 1 & 1 & 1 \\ 0 & 1 & -1 \\ 0 & 0 & 1 \end{bmatrix}.$$

Express this projectivity as the resultant of collineations with centres K_1 and K_0.

EXERCISE 3.2.10 Check that in the 'canonical' relation of duality, $\mathbf{a} \to \mathbf{a}^T\mathbf{x} = 0$, the correlative line k_s of the point K_r is given by

K_r: $r =$	0	1	2	3	4	5	6	7	8	9	10	11	12
k_s: $s =$	1	12	0	6	10	9	3	7	11	2	4	8	5

EXERCISE 3.2.11 Prove that the resultant of the correlation $\mathbf{a} \to \mathbf{a}^T\mathbf{x} = 0$ followed by the correlation $K_r \to k_{-r}$ is the projectivity represented by the matrix

$$\mathbf{M} = \begin{bmatrix} -1 & 0 & 1 \\ 0 & 1 & 0 \\ 1 & 0 & 0 \end{bmatrix}.$$

Prove that $\mathbf{M}$ leaves K_1 invariant, cyclically permutes the points on k_{12} and cyclically permutes the remaining eight points.

EXERCISE 3.2.12 Prove that every Singer matrix of Δ is similar to some power of a multiple of $\mathbf{G}$.

3.3 The Galois plane Φ of order 9

The field $\mathscr{F}$ of order 9 was constructed in § 1.2. $\mathscr{F}$ is commutative and has $\mathscr{D}$ as a subfield. The projective plane over $\mathscr{F}$ we denote by Φ; points are given by vectors $\mathbf{x} = (x, y, z)$ as in § 2.3.

Because of the close resemblance of the field $\mathscr{F}$ to the field of ordinary complex numbers, we shall designate the three elements of $\mathscr{F}$ which lie in $\mathscr{D}$ as the *real elements*, and the six elements of $\mathscr{F} - \mathscr{D}$ as *complex elements*. The term 'complex' we shall always use in this strong sense of 'non-real'; doing this makes possible, for example, the compact and unambiguous statement of Theorem 3.3.1. The complex elements form three conjugate pairs $\pm\epsilon$, $1\pm\epsilon$, $-1\pm\epsilon$, where $\epsilon^2 = -1$; or $\{\omega^r, \omega^{3r}\}$, or $\{\omega^r, \omega^{*r}\}$, $r = \pm 1, 2$ (§ 1.2).

DEFINITION 3.3.1

(i) *Real points*: points (a, b, c), $a, b, c \in \mathscr{D} = \{0, \pm 1\}$.

Real lines: lines $\mathbf{a}^T\mathbf{x} = 0$, $\mathbf{a}$ real (components in $\mathscr{D}$).

Real subplane: Δ_0: the set of 13 real points.

(ii) *Complex points*: the set of points $\Phi - \Delta_0$, namely the 78 points

$$\{(\omega^r, \omega^s, 1), (\omega^r, b, 1), (a, \omega^s, 1), (\omega^r, 1, 0), r, s = 1, 2, 3, 5, 6, 7\}.$$

Complex lines: the 78 lines determined by the corresponding row-vectors $[\omega^r, \omega^s, 1]$, etc.

Conjugate points, P, P^*:

$$P = (\rho, \sigma, \tau) = \mathbf{p}, \quad P^* = (\rho^*, \sigma^*, \tau^*) = \mathbf{p}^*.$$

Conjugate lines: $\mathbf{u}^T\mathbf{x} = 0$, $\mathbf{u}^{*T}\mathbf{x} = 0$.

Since $(\kappa a, \kappa b, \kappa c)$, $\kappa \neq 0$, is the same point as (a, b, c), a real point may be represented by a complex vector; the point with coordinate-vector (ρ, σ, τ) is real if and only if there exists a multiplier $\lambda \neq 0$ such that $\lambda\rho$, $\lambda\sigma$, $\lambda\tau$ are all real. On a real line, say that determined by the points with the real coordinate vectors $\mathbf{a}, \mathbf{b}$, there are four real points, $\mathbf{a}$, $\mathbf{b}$, $\mathbf{a}\pm\mathbf{b}$, and three pairs of conjugate complex points $\{\mathbf{a}+\omega^r\mathbf{b}, \mathbf{a}+\omega^{3r}\mathbf{b}\}$ where $r = \pm 1, 2$.

THEOREM 3.3.1 *Through any complex point P there passes a single real line, namely PP^*. On any complex line $\mathbf{u}^T\mathbf{x} = 0$ there is a single real point, given by $\mathbf{u}^T\mathbf{x} = \mathbf{u}^{*T}\mathbf{x} = 0$.*†

If the point is $(\omega^r, \omega^s, 1)$, the equation of the real line through it is determinable immediately from the appropriate relation of the set

$$\{\omega^2+\omega-1 = 0, \omega^3+\omega+1 = 0, \omega^3-\omega^2-1 = 0\}.$$

If the point is $(a+\epsilon a', b+\epsilon b', c+\epsilon c')$ the real line is

$$\det\begin{bmatrix} x & y & z \\ a & b & c \\ a' & b' & c' \end{bmatrix} = 0.$$

Before we come to any discussion of collineations in general we note that the conjugation operation determines a collineation in Φ:

THEOREM 3.3.2 (i) *The map $\mathscr{J}: P \to P^*$ is a collineation of Φ.* (ii) *$\mathscr{J}$ is not a projectivity.*

Since conjugation is an automorphism of $\mathscr{F}$,

$$\mathbf{u}^T\mathbf{p} = 0 \Leftrightarrow \mathbf{u}^{*T}\mathbf{p}^* = 0,$$

so that $\mathscr{J}$ carries the line $\mathbf{u}^T\mathbf{x} = 0$ to the line $\mathbf{u}^{*T}\mathbf{x} = 0$. $\mathscr{J}$ is not a projectivity because it fixes all the real points, whereas a non-trivial projectivity of Φ cannot fix each vertex of a regular quadrangle.

3.4 Subplanes of Φ

The completion of every regular quadrangle in Φ is a subplane of order 3. Now the Fano plane contains regular quadrangles which, if the Fano plane is to be a subplane of Φ, complete to subplanes of order 3. But the completion of a regular quadrangle in a Fano plane is the Fano plane itself. From this contradiction we deduce:

THEOREM 3.4.1 *There is no subplane of order 2 in Φ.*

† 'On any complex line in Φ there is a single real point' is a special case of Theorem 2.1.6.

From Bruck's theorem we know therefore that the only subplane of order different from 3 is Φ itself.

EXERCISE 3.4.1 Show that Φ contains altogether

$$91.90.81.64/(13.12.9.4) = 7560$$

subplanes of order 3.

It is easily verified that the completion of the regular quadrangle $\boldsymbol{\rho}, \boldsymbol{\sigma}, \boldsymbol{\tau}$, $\boldsymbol{\rho}+\boldsymbol{\sigma}+\boldsymbol{\tau}$ is the subplane $\Delta = \{l\boldsymbol{\rho}+m\boldsymbol{\sigma}+n\boldsymbol{\tau};$ $l, m, n \in \mathscr{D}$ and are not all zero$\}$. That is, $\Delta = \{[\boldsymbol{\rho}, \boldsymbol{\sigma}, \boldsymbol{\tau}]\,(l, m, n)\}$, or say $\{\boldsymbol{\Lambda l}\}$ where $\boldsymbol{\Lambda} = [\boldsymbol{\rho}, \boldsymbol{\sigma}, \boldsymbol{\tau}]$ is a non-singular 3×3 matrix and $\boldsymbol{l}$ is a variable real point. $\boldsymbol{\Lambda}$ is the matrix of one of the projective collineations which maps Δ_0 to Δ. (Given a regular quadrangle in Δ_0 and one in Δ, there is a projectivity which maps the one to the other, and it maps Δ_0 to Δ.)

Let $\boldsymbol{\Lambda} = \mathbf{L}+\epsilon\mathbf{M}$, where $\mathbf{L}$ and $\mathbf{M}$ are real matrices and $\boldsymbol{\Lambda}$ is non-singular, and let $\mathbf{p}$ be a real coordinate-vector. Then $\boldsymbol{\Lambda}\mathbf{p}$ is a real point if and only if, for some $k+\epsilon k' \neq 0$,

$$(\mathbf{L}+\epsilon\mathbf{M})\,(k+\epsilon k')\,\mathbf{p}$$

is a real vector; that is, $(k'\mathbf{L}+k\mathbf{M})\,\mathbf{p} = \mathbf{0}$. The subplane Δ will consist entirely of complex points if this condition is not satisfied for any $\mathbf{p}$, that is:

THEOREM 3.4.2 *If* $\boldsymbol{\Lambda} = \mathbf{L}+\epsilon\mathbf{M}$ *and* $\Delta = \{\boldsymbol{\Lambda}\mathbf{p}$: $\mathbf{p}$ *is a real point*$\}$, *the subplane* Δ *will contain only complex points if and only if none of the matrices* $\mathbf{L}$, $\mathbf{M}$, $\mathbf{L} \pm \mathbf{M}$ *is singular.*

We have, for any ξ, and for some θ, ϕ, depending on $\mathbf{L}, \mathbf{M}$,

$$\det(\mathbf{L}+\xi\mathbf{M}) = \det\mathbf{L}+\theta\xi+\phi\xi^2+\det\mathbf{M}\xi^3,$$

so that if all four matrices $\mathbf{L}$, $\mathbf{M}$, $\mathbf{L} \pm \mathbf{M}$ are singular, then

$$\det\mathbf{L} = \theta = \phi = \det\mathbf{M} = 0$$

and thus $\mathbf{L}+\epsilon\mathbf{M}$ is singular. It follows that at least one of $\mathbf{L}$, $\mathbf{M}$, $\mathbf{L} \pm \mathbf{M}$ is non-singular. Since $\boldsymbol{\Lambda}$ may be replaced by $\boldsymbol{\Lambda}\kappa$ for any $\kappa \neq 0$, we may assume that $\mathbf{L}$ is non-singular. Then $\boldsymbol{\Lambda}\Delta_0 = (\mathbf{1}+\epsilon\mathbf{M}\mathbf{L}^{-1})\,(\mathbf{L}\Delta_0)$. But $\mathbf{L}\Delta_0$ is only a permutation of

Δ_0 and we may therefore always express $\boldsymbol{\Lambda}\Delta_0$ as $(\mathbf{1}+\epsilon\mathrm{N})\,\Delta_0$. Also $\boldsymbol{\Lambda}^{-1} = (\mathbf{1}+\epsilon\mathrm{N})^{-1} = (\mathbf{1}+\mathrm{N}^2)^{-1}(\mathbf{1}-\epsilon\mathrm{N})$, so that the equation of the transform by $\boldsymbol{\Lambda}$ of any line $\mathbf{u}^T\mathbf{x} = 0$ of Δ_0, namely $\mathbf{u}^T\boldsymbol{\Lambda}^{-1}\mathbf{x} = 0$, may be written as $\{\mathbf{u}^T(\mathbf{1}+\mathrm{N}^2)^{-1}\}(\mathbf{1}-\epsilon\mathrm{N})\,\mathbf{x} = 0$ or

$$\mathbf{v}^T(\mathbf{1}-\epsilon\mathrm{N})\,\mathbf{x} = 0.$$

The two subplanes Δ_0 and Δ contain no common line† if and only if there is no vector $\mathbf{v}$ and scalar κ such that $\mathbf{v}^T(\mathbf{1}-\epsilon\mathrm{N})\kappa$ is real, the condition for which is that the three matrices N, $\mathbf{1}\pm\mathrm{N}$ are all non-singular. This is precisely the condition that Δ_0 and Δ have no common point, so that:

Lemma 1 *Two subplanes of order* 3 *in* Φ *have no common line if and only if they have no common point.*

Before proceeding to the main theorems, we state a property of matrices which plays an important part in the proof.

Exercise 3.4.2 Prove that every 3×3 matrix is similar to its transpose.

Theorem 3.4.3 *The common points and lines of any two subplanes of order* 3 *in* Φ *form a self-dual configuration.*

Theorem 3.4.4 *Two subplanes of order* 3 *in* Φ *either are disjoint or they have common:*

(i) *one point and one line either disjoint or incident,*

(ii) *two points, the line joining them and another line incident with one of them,*

(iii) *three non-collinear points and the lines joining the pairs of them,*

(iv) *four collinear points, the line containing them and three other lines incident with one of them, or*

(v) *five points, four of them collinear, the line containing the four points, and the four lines joining these four to the fifth point.*

† The statement 'Δ contains l' is used as the equivalent of '$\Delta \cap l = \{4 \text{ points}\}$', so that '$\Delta_0$ and Δ have a common line' is equivalent to '$\Delta_0 \cap l = \{4 \text{ points}\}$ and $\Delta \cap l = \{4 \text{ points}\}$'. *A priori* the two sets of four points may have common 0, 1, 2, 3 or 4 members; we shall prove shortly that in Φ the only case which does not occur is that of (exactly) three collinear common points.

Since any subplane may be taken to be Δ_0, and every other subplane is included in the set

$$\{(\mathbf{1}+\epsilon\mathrm{N})\,\Delta_0 : \mathrm{N} \text{ real}, \ \mathbf{1}+\epsilon\mathrm{N} \text{ non-singular}\},$$

the determination of the intersection of two subplanes can be made to depend on the properties of $\mathbf{N}$. Assume that $\Delta_0 \cap \Delta$ contains at least one point $\mathbf{p}$, take $\Delta = \mathbf{\Lambda}'\Delta_0$ and let $\kappa\mathbf{p} = \mathbf{\Lambda}'\mathbf{q}$ (that is, $\mathbf{q}$ is the point of Δ_0 from which the real point $\mathbf{p}$ in Δ is derived). Let $\mathbf{H}$ be any real non-singular matrix such that $\mathbf{Hp} = \mathbf{q}$, and write $\mathbf{\Lambda} = \mathbf{\Lambda}'\mathbf{H}$, then

$$\mathbf{\Lambda}\Delta_0 = \mathbf{\Lambda}'(\mathbf{H}\Delta_0) = \mathbf{\Lambda}'\Delta_0 = \Delta,$$

and $$\mathbf{\Lambda}\mathbf{p} = \mathbf{\Lambda}'(\mathbf{Hp}) = \mathbf{\Lambda}'\mathbf{q} = \kappa\mathbf{p}.$$

Thus $\mathbf{\Lambda}$ determines a collineation $\mathscr{L}$ of Φ which maps Δ_0 to Δ and fixes $\mathbf{p}$ in $\Delta_0 \cap \Delta$.

Now take the point $\mathbf{p}$ to be $\mathbf{e}_1 = (1, 0, 0)$, so that $\mathbf{\Lambda}\mathbf{e}_1 = \kappa\mathbf{e}_1$ and the first column of $\mathbf{\Lambda}$ is $\kappa\mathbf{e}_1$ with $\kappa \neq 0$. Since non-singular matrices in which the first column is a scalar multiple of $\mathbf{e}_1$ form a group under multiplication, we may apply the argument which led to lemma 1 and reduce $\mathbf{\Lambda}$ to the form $\mathbf{1}+\epsilon\mathbf{N}$, where

$$\mathrm{N} = [a\mathbf{e}_1, \mathbf{b}, \mathbf{c}].$$

Let $\mathbf{r}$ be any point of $\Delta_0 \cap \Delta$, so that for some $\mathbf{r}' \in \Delta_0$ and some $k, k' \in \mathscr{D}$, $(\mathbf{1}+\epsilon\mathbf{N})\,\mathbf{r} = (1+\epsilon k)\,k'\mathbf{r}'$. Then $k'\mathbf{r}' = \mathbf{r}$ and every common point $\mathbf{r}$ is such that, for some k, $\mathbf{Nr} = k\mathbf{r}$. We now make use of the theorem in Exercise 3.4.2, namely, that we can find a real matrix $\mathbf{B}$ such that $\mathbf{N}^T = \mathbf{BNB}^{-1}$, and that, consequently, $\mathbf{\Lambda}^T = \mathbf{B\Lambda B}^{-1}$, and

$$\mathbf{\Lambda r} = \kappa\mathbf{r} \Rightarrow \mathbf{r}^T\mathbf{\Lambda}^T = \kappa\mathbf{r}^T \Rightarrow (\mathbf{r}^T\mathbf{B})\,\mathbf{\Lambda} = \kappa(\mathbf{r}^T\mathbf{B}).$$

Under the collineation $\mathscr{L}$ we have

$$\mathscr{L}\colon (\mathbf{r}^T\mathbf{B})\,\mathbf{x} = 0 \to (\mathbf{r}^T\mathbf{B})\,\mathbf{\Lambda}^{-1}\mathbf{x} = 0 \equiv \kappa^{-1}(\mathbf{r}^T\mathbf{B})\,\mathbf{x} = 0,$$

so that, paired with each point $\mathbf{r}$ of $\Delta_0 \cap \Delta$, there is a common line, $\mathbf{r}^T\mathbf{Bx} = 0$, and the configuration of common lines is obtained from that of the common points by the correlation

$$\mathbf{r} \to \mathbf{r}^T\mathbf{Bx} = 0. \quad \text{(Theorem 3.4.3.)}$$

For the common point $\mathbf{e}_1$ we have, writing $\alpha = 1+a\epsilon$,

$$(\mathbf{1}+\epsilon\mathbf{N})\,\mathbf{e}_1 = \alpha\mathbf{e}_1;$$

consequently, α is the eigenvalue of $\mathbf{\Lambda}$ corresponding to the right-eigenvector $\mathbf{e}_1$, and there exist therefore left-eigenvectors, $\mathbf{u}$, such that $\mathbf{u}^T\mathbf{\Lambda} = \mathbf{u}^T(\mathbf{1}+\epsilon\mathbf{N}) = \alpha\mathbf{u}^T$. For such a vector

$$\mathscr{L}\colon \alpha\mathbf{u}^T\mathbf{x} = 0 \equiv \mathbf{u}^T\mathbf{\Lambda}\mathbf{x} = 0 \rightarrow (\mathbf{u}^T\mathbf{\Lambda})\,\mathbf{\Lambda}^{-1}\mathbf{x} = 0 \equiv \mathbf{u}^T\mathbf{x} = 0,$$

so that $\mathbf{u}^T\mathbf{x} = 0$ is a line common to Δ_0 and Δ. The vector $\mathbf{u}$ satisfies the two equations

$$u_1 b_1 + u_2(b_2 - a) + u_3 b_3 = 0,$$

$$u_1 c_1 + u_2 c_2 + u_3(c_3 - a) = 0.$$

If $\mathbf{e}_1$ is the only common point, then $\mathbf{u}^T\mathbf{x} = 0$ is the only common line, and it passes through $\mathbf{e}_1$ if and only if $u_1 = 0$, that is

$$(b_2 - a)\,(c_3 - a) = b_3 c_2.$$

Thus, if Δ_0 and Δ have a single common point, and consequently a single common line, the point and line may be either disjoint or incident.

Suppose now there is a common point $\mathbf{r} = (r, s, t) \neq \mathbf{e}_1$. From the relation $\mathbf{N}\mathbf{r} = k\mathbf{r}$ above we have

$$(a-k)\,r + b_1 s + c_1 t = 0,$$

$$(b_2 - k)\,s + c_2 t = 0,$$

$$b_3 s + (c_3 - k)\,t = 0.$$

From the second and third equations, we obtain

$$b_3 s^2 + (c_3 - b_2)\,st - c_2 t^2 = 0.$$

Assume that $a \neq k$, and that at least one of the coefficients in the two equations is not zero, then, if a pair $(s, t) \neq (0, 0)$ satisfies the equations, r and k are uniquely determined by (s, t). Thus, according as $(c_3 - b_2)^2 + b_3 c_2 = -1$ or 0 or 1, under these assumptions the number of common points additional to $\mathbf{e}_1$ is 0 or 1 or 2.

EXERCISE 3.4.3 Discuss the case $a = k$.

Finally, let all the coefficients in the two equations be zero, that is, $c_3 = b_2 = k$, $c_2 = b_3 = 0$, so that, since $\mathbf{N} \neq k\mathbf{1}$, at least one of the coefficients $a-k$, b_1, c_1 is not zero. In this case all four points of Δ_0 that lie on the line $(a-k)x+b_1y+c_1z = 0$ also lie in Δ. If $a \neq k$, this line is disjoint from $\mathbf{e}_1$ giving four additional points, but, if $a = k$, then $\mathbf{e}_1$ is one of the four points on the line.

The only case presenting any further problem is that of exactly two points additional to $\mathbf{e}_1$, which, *a priori*, might be collinear with $\mathbf{e}_1$. But, if they are collinear with $\mathbf{e}_1$, we may take them to be (r, s, t) and (r', s, t) with $r \neq r'$; this is impossible, since unless there is a fourth point in the set of collinear common points, the pair (s, t) uniquely determines the coordinate r. Thus, if $\Delta \cap \Delta_0$ consists of exactly three points, these points are non-collinear. This completes the proof of Theorem 3.4.4.

EXERCISE 3.4.4 Find, for each pattern of common points, a matrix $\mathbf{N}$ which has only one, two or three non-zero elements.

3.5 Sets of mutually disjoint subplanes

In this section we are to adapt the cyclic Table 3.1.2 of incidences in the real subplane Δ_0 to provide a simple incidence scheme (Table 3.5.2) for the points and lines of Φ, which later we shall be able to convert into a scheme for the miniquaternion plane Ψ. The geometric property of Φ which is an immediate consequence of the structure of the incidence table is the following:

THEOREM 3.5.1 *The* 91 *points of* Φ *can be split into seven mutually disjoint subplanes of order* 3. *Any line of* Φ *contains four points of one of these subplanes and one point in each of the other six.*

Let
$$\mathbf{G} = \begin{bmatrix} 0 & 0 & 1 \\ 1 & 0 & 1 \\ 0 & 1 & 0 \end{bmatrix},$$

as in §3.2. Let $\mathbf{p} \in \Delta_0$ and let the three points $\mathbf{p}$, $\mathbf{G}^r\mathbf{p}$, $\mathbf{G}^s\mathbf{p}$ of the Singer cycle generated by $\mathbf{G}$ in Δ_0 be collinear. Then the points of the sets $\{\mathbf{G}^t\mathbf{p}, \mathbf{G}^{t+r}\mathbf{p}, \mathbf{G}^{t+s}\mathbf{p}\}$, $t = 0, \ldots, 12$, are also collinear. (We could choose, say, $\{r, s\} \subset \{1, 3, 9\}$. (See Exercise 3.2.4.) Any point of Φ has a coordinate-vector $\mathbf{p}+\epsilon\mathbf{q}$, where $\mathbf{p}$, $\mathbf{q}$ are real.

Since $\mathbf{p}$, $\mathbf{q} \in \Delta_0$, for some index c and some constant $k \in \{\pm 1\}$, $\mathbf{q} = \mathbf{G}^c(k\mathbf{p})$. For any $\mathbf{p}$, the set of points $\{\mathbf{p}, \mathbf{G}^r\mathbf{p}, \mathbf{G}^s\mathbf{p}\}$ is collinear, so that there are real multipliers l, m, n, not all zero, such that $l\mathbf{p} + m\mathbf{G}^r\mathbf{p} + n\mathbf{G}^s\mathbf{p} = \mathbf{0}$, and therefore also

$$l\mathbf{q} + m\mathbf{G}^r\mathbf{q} + n\mathbf{G}^s\mathbf{q} = k\mathbf{G}^c(l\mathbf{p} + m\mathbf{G}^r\mathbf{p} + n\mathbf{G}^s\mathbf{p}) = \mathbf{0}.$$

Thus $$l(\mathbf{p} + \epsilon\mathbf{q}) + m\mathbf{G}^r(\mathbf{p} + \epsilon\mathbf{q}) + n\mathbf{G}^s(p + \epsilon\mathbf{q}) = \mathbf{0},$$

and among the points of the set $\{\mathbf{G}^t(\mathbf{p} + \epsilon\mathbf{q}); t = 0, \ldots, 12\}$ there are the same relations of collinearity as among those of the set $\{\mathbf{G}^t\mathbf{p}\}$. Hence the thirteen points $\mathbf{G}^t(\mathbf{p} + \epsilon\mathbf{q})$ form a subplane Δ. The sets derived in this way from two complex points either are identical or they are disjoint, so that we can construct successively, as Singer cycles generated by $\mathbf{G}$, altogether seven mutually disjoint subplanes Δ.

A set of seven such subplanes may be generated explicitly thus: on $z = 0$ there are the four real points $(1, 0, 0)$, $(\pm 1, 1, 0)$, $(0, 1, 0)$, and three conjugate complex pairs, say:

$$A_0 = (\omega, 1, 0), \qquad B_0 = (-\omega^2, 1, 0), \qquad C_0 = (-\omega^3, 1, 0),$$
$$A_0' = (\omega^3, 1, 0), \qquad B_0' = (\omega^2, 1, 0), \qquad C_0' = (-\omega, 1, 0).$$

(The naming of the points of a pair may at the outset be made arbitrarily; the choice above is dictated by subsequent developments.) From the points A_0 we construct the sequence of points $A_r = \mathbf{G}^r A_0$, namely

$A_r: r =$	0	1	2	3	4	5	6	7	8	9	10	11	12
Coord.	ω	0	1	ω	1	$-\omega^3$	$-\omega^3$	$-\omega^2$	ω^3	$-\omega$	1	$-\omega^2$	ω^2
Vector	1	ω	1	$-\omega^3$	$-\omega^3$	$-\omega^2$	ω^3	$-\omega$	1	$-\omega^2$	ω^2	ω	0
	0	1	ω	1	$-\omega^3$	$-\omega^3$	$-\omega^2$	ω^3	$-\omega$	1	$-\omega^2$	ω^2	ω

We determine first the equations of the joins, $A_r X$, of these points to $X = (1, 0, 0)$.

$$A_0 \in z = 0, \quad A_4 \in z = y, \quad A_{10} \in z = -y, \quad A_{12} \in y = 0,$$

$$\{A_2, A_3, A_5, A_{11}\} \subset z = \omega y,$$

$$A_1 \in z = -\omega^3 y, \quad A_6 \in z = \omega^3 y, \quad A_7 \in z = -\omega^2 y, \quad A_8 \in z = -\omega y,$$

$$A_9 \in z = \omega^2 y.$$

Table 3.5.1 *Points K_r, A_r, B_r, C_r, lines $A_r X$, etc.*

r	Line k_r in Δ_0	Δ_0	Δ_A		Δ_B		Δ_C	
		$K_r = \mathbf{G}^r X$	$A_r = \mathbf{G}^r A_0$	$A_r X$	$B_r = \mathbf{G}^r B_0$	$B_r X$	$C_r = \mathbf{G}^r C_0$	$C_r X$
0	0 0 1	1 0 0	ω 1 0	k_0	$-\omega^2$ 1 0	k_0	$-\omega^3$ 1 0	k_0
1	1 0 0	0 1 0	0 ω 1	c_0	0 $-\omega^2$ 1	b_0	0 $-\omega^3$ 1	a_0
2	−1 1 0	0 0 1	1 1 ω	a_0	1 1 $-\omega^2$	b_0'	1 1 $-\omega^3$	c_0
3	1 −1 1	1 1 0	ω $-\omega^3$ 1	a_0	$-\omega^2$ ω 1	c_0	$-\omega^3$ $-\omega^2$ 1	b_0
4	0 −1 1	0 1 1	1 $-\omega^3$ $-\omega^3$	k_4	1 ω ω	k_4	1 $-\omega^2$ $-\omega^2$	k_4
5	−1 0 1	1 1 1	$-\omega^3$ $-\omega^2$ $-\omega^3$	a_0	ω $-\omega^3$ ω	b_0	$-\omega^2$ ω $-\omega^2$	c_0'
6	1 1 0	1 −1 1	$-\omega^3$ ω^3 $-\omega^2$	a_0'	ω $-\omega$ $-\omega^3$	b_0	$-\omega^2$ ω^2 ω	c_0
7	−1 1 1	1 −1 −1	$-\omega^2$ $-\omega$ ω^3	b_0'	$-\omega^3$ ω^2 $-\omega$	a_0'	ω ω^3 ω^2	c_0
8	−1 −1 1	−1 0 −1	ω^3 1 $-\omega$	c_0'	$-\omega$ 1 ω^2	b_0	ω^2 1 ω^3	a_0'
9	1 1 1	−1 1 0	$-\omega$ $-\omega^2$ 1	b_0	ω^2 $-\omega^3$ 1	a_0	ω^3 ω 1	c_0
10	0 1 1	0 −1 1	1 ω^2 $-\omega^2$	k_{10}	1 ω^3 $-\omega^3$	k_{10}	1 $-\omega$ ω	k_{10}
11	1 0 1	1 1 −1	$-\omega^2$ ω ω^2	a_0	$-\omega^3$ $-\omega^2$ ω^3	c_0'	ω $-\omega^3$ $-\omega$	b_0'
12	0 1 0	−1 0 1	ω^2 0 ω	k_{12}	ω^3 0 $-\omega^2$	k_{12}	$-\omega$ 0 $-\omega^3$	k_{12}

$$a_0 : z = \omega y, \quad b_0 : z = \omega^2 y, \quad c_0 : z = -\omega^3 y$$

$$a_0' : z = \omega^3 y, \quad b_0' : z = -\omega^2 y, \quad c_0' : z = -\omega y$$

Let us denote the line through X containing the four points A_2, A_3, A_5, A_{11} by a_0, and allocate names consistently to the other lines through X, so that, for example, b_0 contains four points B_r.

Since the coordinate-vectors of a pair of points such as $\{A_r, A_r'\}$ are conjugate complexes we need to list only one of them. The points in the four subplanes Δ_0, $\Delta_A = \{A_r\}$, Δ_B, Δ_C, the lines in Δ_0, and the lines $A_r X$, etc., are set out in the Table 3.5.1.

Exercise 3.5.1 Construct the pair of columns for $\Delta_{A'}$. Check that $A_r' \in k_r$ for each r.

By rearranging this table to show the sets of ten points lying on the six lines $A_r X$, etc., we obtain Table 3.5.2.

Table 3.5.2

$k_r = \{K_r, K_{r+1}, K_{r+3}, K_{r+9}, A_r, B_r, C_r, A_r', B_r', C_r'\}$

	Points						
Line	K_r	A_r	A_r'	B_r	B_r'	C_r	C_r'
a_0	0	2, 3, 5, 11	6	9	7	1	8
a_0'	0	6	2, 3, 5, 11	7	9	8	1
b_0	0	9	7	1, 5, 6, 8	2	3	11
b_0'	0	7	9	2	1, 5, 6, 8	11	3
c_0	0	1	8	3	11	2, 6, 7, 9	5
c_0'	0	8	1	11	3	5	2, 6, 7, 9

Using the fact that this table is derived from the Singer cycles, we can state the rules for determining from the table the points on every line in Φ. Thus,

let $P \in \{K, A, B, C, A', B', C'\}$ and $u \in \{k, a, b, c, a', b', c'\}$,

then $$P_r \in u_0 \Rightarrow P_{r+s} \in u_s \quad (s = 0, 1, \ldots, 12),$$

where the indices are reduced modulo 13. Since the points and lines together form a projective plane, we should be able to identify from the tables the line joining any two points and the point common to any two lines. The various types of pairs of points are

$A_r K_s$: $A_{r-s} K_0$ is a line listed in the Table, say u_0.

$$A_{r-s} K_0 = u_0 \Rightarrow A_r K_s = u_s.$$

$A_r B_r = k_r$.

$A_r A_s$: The difference $s-r$ mod 13 occurs exactly once among the differences of pairs in the set of numbers $\{2, 3, 5, 11\}$, (say as the difference of $s+t$, $r+t$). Then

$$A_{r+t}A_{s+t} = u_0 \Rightarrow A_r A_s = u_{-t}.$$

$A_r B_s$: The difference $s-r$ occurs exactly once among the differences: $6-\{2,3,5,11\}, \{2,3,5,11\}-6, 9-7, 7-9, 8-1, 1-8$. $A_{r+t}B_{s+t} \in u_0 \Rightarrow A_r B_s \in u_{-t}$.

The common point of two lines can be determined similarly.

Exercise 3.5.2 Find the lines A_4A_{10}, $A_4B'_8$, A_4B_9, A_4K_3, and the points $a_3 \cap a_5$, $a_3 \cap c'_{12}$.

Exercise 3.5.3 Find the lines and points in the subplane defined by the regular quadrangle $K_0A_1B_2C_4$.

The essential property of Table 3.5.2, which ensures that joins and intersections are uniquely defined, is that the differences, modulo 13, between indices in one column and indices in the same rows in another column should constitute the set $\{1, 2, \ldots, 12\}$. For example, the differences for the columns A_r, B'_r (numbers in B'_r subtracted from numbers in A_r) are

$$(8, 9, 11, 4), 10, 7, (6, 2, 1, 12), 3, 5.$$

3.6 Singer cycles in Φ

The theorems on Singer cycles in Δ (§3.2) have their counterparts in Φ, but, as might be expected, are more complicated in their detail. Every element κ of Φ is such that $\kappa^9 = \kappa$ (since $\omega^8 = 1$), and since $\det(\rho\mathbf{M}) = \rho^3 \det \mathbf{M}$, if $\det \mathbf{M} = m \neq 0$ then

$$\det(m^5\mathbf{M}) = 1.$$

We assume, then, that if there is a Singer cycle it is generated by a matrix $\mathbf{S}$ such that $\det \mathbf{S} = 1$. In order to obtain a Singer cycle we require that $\chi_1 = \chi(\mathbf{S}) = \det(x\mathbf{1} - \mathbf{S}) = x^3 + ux^2 + vx - 1$ is irreducible. Let us start with say

$$\mathbf{S} = \begin{bmatrix} 0 & 0 & 1 \\ 1 & 0 & \omega^2 \\ 0 & 1 & 0 \end{bmatrix}, \quad \chi_1 = x^3 - \omega^2 x - 1.$$

Now calculate $\chi_2, \chi_4, \ldots$ by applying Theorem 2.5.4 (i), namely

$$\chi_r \equiv x^3+u_r x^2+v_r x-1 \Rightarrow \chi_{2r} \equiv x^3-(u_r^2+v_r)\,x^2-(u_r-v_r^2)\,x-1.$$

That is, $u_{2r} = -u_r^2-v_r$, $v_{2r} = -u_r+v_r^2$, so that, listing by pairs of values u_r, v_r, we obtain the following sequence of characteristic functions $\chi_1, \chi_2, \chi_4, \ldots$:

$$\chi_r = x^3+u_r x^2+v_r x-1.$$

$r =$	1	2	4	8	16	32	64	37	74	57	23	46	(1)
$u_r =$	0	ω^2	-1	ω^3	-1	$-\omega^2$	$-\omega^2$	1	$-\omega^3$	$-\omega$	ω^3	$-\omega$	(0)
$v_r =$	$-\omega^2$	-1	ω	ω^3	$-\omega$	ω^3	0	ω^2	1	$-\omega$	1	$-\omega^2$	$(-\omega^2)$

From this set of powers of $\mathbf{S}$ we find two other sets with the same sequence of functions χ by using the relation $\chi_1 = \chi_9 = \chi_{81}$ (Theorem 2.5.4 (iii)).

There will be a corresponding set of three sequences of 12 functions for which u_r, v_r take the conjugate complex values; we seek therefore a matrix $\mathbf{S}^r$ with $\chi(\mathbf{S}^r) = x^3+\omega^2 x-1$, and find, as we might have expected, that

$$\mathbf{S}^3 = \begin{bmatrix} 1 & 0 & \omega^2 \\ \omega^2 & 1 & -1 \\ 0 & \omega^2 & 1 \end{bmatrix}$$

satisfies the condition.

We have now six disjoint sets of 12, and these account for 72 out of the 90 powers $\mathbf{S}^r$, $r = 1, \ldots, 90$. Of the remaining 18, 12 are the matrices $\mathbf{S}^{7r}$ and the other 6 are $\mathbf{S}^{13r}$. The complete set of powers of $\mathbf{S}$ obtained by successive squarings and of the functions χ_r is listed in Table 3.6.1.

The table shows 30 different functions x^3+ux^2+vx-1, and we are to prove that there are no other irreducible polynomials. We do so by counting the reducible polynomials. There are 81 polynomials x^3+ux^2+vx-1. The reducible polynomials are of two types: (i) three linear factors; (ii) one linear factor and one irreducible quadratic factor.

(i) $(x-\omega^r)(x-\omega^s)(x-\omega^t)$ with $\omega^{r+s+t} = 1$, so that

$$r+s+t = 0 \text{ or } 8 \text{ or } 16 \quad (r, s, t = 0, \ldots, 7).$$

Table 3.6.1 *Irreducible functions* $\chi_r \equiv x^3 + u_r x^2 + v_r x - 1$

					Conjugates					$\mathbf{S}^{7r}$				
u_r	v_r	r	$9r$	$81r$	u_r	v_r	r	$9r$	$81r$	u_r	v_r	r	$9r$	$81r$
0	$-\omega^2$	1	9	81	0	ω^2	3	27	61	1	1	7	63	21
ω^2	-1	2	18	71	$-\omega^2$	-1	6	54	31	1	0	14	35	42
-1	ω	4	36	51	-1	ω^3	12	17	62	-1	-1	28	70	84
ω^3	ω^3	8	72	11	ω	ω	24	34	33	0	-1	56	49	77
-1	$-\omega$	16	53	22	-1	$-\omega^3$	48	68	66	.	.	.	.	.
$-\omega^2$	ω^3	32	15	44	ω^2	ω	5	45	41					
$-\omega^2$	0	64	30	88	ω^2	0	10	90	82			$\mathbf{S}^{13r}$		
1	ω^2	37	60	85	1	$-\omega^2$	20	89	73					
$-\omega^3$	1	74	29	79	$-\omega$	1	40	87	55	$-\omega^3$	ω	13	26	52
$-\omega$	$-\omega$	57	58	67	$-\omega^3$	$-\omega^3$	80	83	19	.	.	.	.	.
ω^3	1	23	25	43	ω	1	69	75	38	$-\omega$	ω^3	39	78	65
$-\omega$	$-\omega^2$	46	50	86	$-\omega^3$	ω^2	47	59	76	.	.	.	.	.
										.	.	.	.	.

There is no simple formula for the number of such triplets r, s, t, but we can list them; arrange them so that $r \geqslant s \geqslant t$.

$r =$	0	7	6	6	5	5	4	4	4	3	7	7	7	6	6
$s =$	0	1	2	1	3	2	4	3	2	3	7	6	5	6	5
$t =$	0	0	0	1	0	1	0	1	2	2	2	3	4	4	5

There are thus 15 altogether.

(ii) $(x-\omega^{-s})(x^2+\kappa x+\omega^s) \quad (s = 0, 1, \ldots, 7;\ \kappa = 0, 1, \omega, \ldots, \omega^7)$.

There are 72 of these functions, and we have to find the number in which the quadratic factor is irreducible. The reducible quadratic polynomials are $(x-\omega^l)(x-\omega^m)$ $(l, m = 0, 1, \ldots, 7)$ and there are just 36 of these. There are therefore also 36 irreducible quadratics, and altogether 51 reducible cubics; there remain therefore 30 irreducible cubics. It follows that:

THEOREM 3.6.1 *Every Singer matrix of Φ is similar to some power of $\rho\mathbf{S}$, where*

$$\mathbf{S} = \begin{bmatrix} 0 & 0 & 1 \\ 1 & 0 & \omega^2 \\ 0 & 1 & 0 \end{bmatrix}.$$

The complete set of points generated as $\mathbf{S}^r(1, 0, 0)$ is shown in Table 3.6.2.

Again $\mathbf{S}^r$ is the matrix whose three columns are those numbered $r, r+1, r+2$. The points on the line $z = 0$ occur in the columns numbered:

0	1	3	9	27	81	61	49	77	56
0	1	3	3^2	3^3	3^4	3^5	49	49×9	49×9^2

so that we may take this to be the first column of the cyclic table of columns $(r, r+1, r+3, \ldots, r+56)$, $r = 0, \ldots, 90$, which lists the sets of ten collinear points in Φ. We extract from the whole cyclic table the columns which contain the point 0, rearranging the rows so that in the first column the first four entries are multiples of 7 and the remaining six are in the order of the remainders modulo 7.

Table 3.6.2 Φ: *points* $\mathbf{S}^r\begin{bmatrix}1\\0\\0\end{bmatrix}$, $\mathbf{S} = \begin{bmatrix}0 & 0 & 1\\1 & 0 & \omega^2\\0 & 1 & 0\end{bmatrix}$

r	0	1	2	3	4	5	6	7	8	9	10	11	12	13	14	15	16	17	18
x	1	0	0	1	0	ω^2	1	-1	$-\omega^2$	ω	0	1	ω	ω^2	$-\omega$	$-\omega^2$	-1	ω^2	ω^2
y	0	1	0	ω^2	1	-1	$-\omega^2$	ω	0	1	ω	ω^2	$-\omega$	$-\omega^2$	-1	ω^2	ω^2	1	$-\omega$
z	0	0	1	0	ω^2	1	-1	$-\omega^2$	ω	0	1	ω	ω^2	$-\omega$	$-\omega^2$	-1	ω^2	ω^2	1
r	19	20	21	22	23	24	25	26	27	28	29	30	31	32	33	34	35	36	37
x	1	$-\omega$	$-\omega^2$	$-\omega^2$	ω^2	ω	$-\omega^3$	ω	$-\omega$	0	ω^2	$-\omega$	-1	-1	-1	$-\omega^3$	$-\omega^3$	$-\omega^2$	ω^2
y	$-\omega^2$	$-\omega^2$	ω^2	ω	$-\omega^3$	ω	$-\omega$	0	ω^2	$-\omega$	-1	-1	-1	$-\omega^3$	$-\omega^3$	$-\omega^2$	ω^2	$-\omega^2$	$-\omega^3$
z	$-\omega$	$-\omega^2$	$-\omega^2$	ω^2	ω	$-\omega^3$	ω	$-\omega$	0	ω^2	$-\omega$	-1	-1	-1	$-\omega^3$	$-\omega^3$	$-\omega^2$	ω^2	$-\omega^2$
r	38	39	40	41	42	43	44	45	46	47	48	49	50	51	52	53	54	55	56
x	$-\omega^2$	$-\omega^3$	ω^3	ω^3	1	$-\omega^2$	ω	-1	1	ω^3	$-\omega$	ω^2	0	ω^3	ω^2	$-\omega$	ω^2	-1	ω^3
y	ω^3	ω^3	1	$-\omega^2$	ω	-1	1	ω^3	$-\omega$	ω^2	0	ω^3	ω^2	$-\omega$	ω^2	-1	ω^3	0	ω^3
z	$-\omega^3$	ω^3	ω^3	1	$-\omega^2$	ω	-1	1	ω^3	$-\omega$	ω^2	0	ω^3	ω^2	$-\omega$	ω^2	-1	ω^3	0
r	57	58	59	60	61	62	63	64	65	66	67	68	69	70	71	72	73	74	75
x	0	ω^3	ω^3	$-\omega$	$-\omega^2$	0	ω^2	$-\omega^2$	-1	ω^3	ω^2	ω^3	ω^2	$-\omega^3$	ω^2	1	ω	$-\omega^2$	$-\omega$
y	ω^3	$-\omega$	$-\omega^2$	0	ω^2	$-\omega^2$	-1	ω^3	ω^2	ω^3	ω^2	$-\omega^3$	ω^2	1	ω	$-\omega^2$	$-\omega$	$-\omega^3$	$-\omega$
z	ω^3	ω^3	$-\omega$	$-\omega^2$	0	ω^2	$-\omega^2$	-1	ω^3	ω^2	ω^3	ω^2	$-\omega^3$	ω^2	1	ω	$-\omega^2$	$-\omega$	$-\omega^3$
r	76	77	78	79	80	81	82	83	84	85	86	87	88	89	90	(0)	(1)		
x	$-\omega^3$	$-\omega$	0	ω^3	$-\omega$	$-\omega$	0	1	$-\omega$	ω^2	$-\omega^2$	ω^3	ω^3	-1	$-\omega^2$	1	0		
y	0	ω^3	$-\omega$	$-\omega$	0	1	$-\omega$	ω^2	$-\omega^2$	ω^3	ω^3	-1	$-\omega^2$	1	0	0	1		
z	$-\omega$	0	ω^3	$-\omega$	$-\omega$	0	1	$-\omega$	ω^2	$-\omega^2$	ω^3	ω^3	-1	$-\omega^2$	1	0	0		

Table 3.6.3 *Key collinear sets* $\{\mathbf{S}^r(1,0,0)\}$ *in* Φ (*read by columns*)

$r =$									
0	42	35	14	90	82	88	10	30	[64]
49	0	84	63	48	40	46	59	79	[22]
56	7	0	70	55	47	53	66	86	[29]
77	28	21	0	76	68	74	87	16	[50]
[1]	[43]	[36]	[15]	0	83	89	11	31	65
9	51	44	23	[8]	0	6	19	39	73
3	45	38	17	2	[85]	0	13	33	67
81	32	25	4	80	72	[78]	0	20	54
61	12	5	75	60	52	58	[71]	0	34
27	69	62	41	26	18	24	37	[57]	0

The multiples of 7 form a subtable

$7\times$

0	6	5	2
7	0	12	9
8	1	0	10
11	4	3	0

which is part of a cyclic table for the collinear sets of a plane of order 3. That is, the thirteen points of the set $\{7r\colon r = 0, \ldots, 12\}$ form a subplane Δ, and therefore also each of the sets $\{7r+1\}$ (marked $[m]$ in Table 3.6.3), $\{7r+2\}, \ldots, \{7r+6\}$ forms a subplane Δ. The table thus displays Φ as the union of seven mutually disjoint subplanes Δ_s, where Δ_s is the set of thirteen points $\{7r+s\colon r = 0, \ldots, 12\}$, $s = 0, \ldots, 6$.

Embedded in the plane Φ is the real subplane Δ_0; in Table 3.6.2 the real points are numbered

$$0 = K_0,\;\; 1 = K_1,\;\; 2 = K_2,\;\; 21 = (1,-1,1) = K_6,$$

$$26 = (1,0,-1) = K_{12},\quad 31 = (1,1,1) = K_5,$$

$$36 = (1,1,-1) = K_{11},\quad 39 = (-1,1,1) = K_7,$$

$$56 = (1,1,0) = K_3,\quad 57 = (0,1,1) = K_4,$$

$$61 = (-1,1,0) = K_9,\quad 62 = (0,-1,1) = K_{10},$$

$$80 = (1,0,1) = K_8.$$

EXERCISE 3.6.1 It is to be proved that the matrix

$$\mathbf{W} = \omega\mathbf{1}+\mathbf{G} = \begin{bmatrix} \omega & 0 & 1 \\ 1 & \omega & 1 \\ 0 & 1 & \omega \end{bmatrix}$$

generates a Singer cycle in Φ in which the points K_r of Δ_0 occur in the positions numbered $7s$ (but in a different order), the points A_r occur in positions $7s+1$, etc.

(i) Verify that the matrix $-\omega\mathbf{W}$ has the same characteristic polynomial as the matrix $\mathbf{S}$ (Table 3.6.2). $-\omega\mathbf{W}$ is therefore similar to $\mathbf{S}$ and generates the same table of key collinear sets. It follows that $\mathbf{W}$ also generates the same table.

(ii) Verify that $\mathbf{W}^7 = -\omega\mathbf{G}^6$ and

$$\begin{matrix} \mathbf{W}K_0 & \mathbf{W}^2K_0 & \mathbf{W}^3K_0 & \mathbf{W}^4K_0 & \mathbf{W}^5K_0 & \mathbf{W}^6K_0 & \mathbf{W}^7K_0 \\ = A_0 & B_7 & C'_0 & C_{12} & B'_4 & A'_8 & K_6 \end{matrix}$$

so that we obtain the Table 3.6.4.

Table 3.6.4 *Position numbers of* $K, A, \ldots$ *in the Singer cycle* $\mathbf{W}^t(1,0,0)$

$t =$	$7r$	$7r+1$	$7r+2$	$7r+3$	$7r+4$	$7r+5$	$7r+6$	$r = 0, \ldots, 12$
Point	K_{6r}	A_{6r}	B_{6r+7}	C'_{6r}	C_{6r-1}	B'_{6r+4}	A'_{6r+8}	(mod 13)

EXERCISE 3.6.2 Verify that the table corresponding to Table 3.6.3, using the symbols $K_r, A_r, \ldots$ instead of the position-numbers in the Singer cycle, is the following:

Table 3.6.5 *Key collinear sets in* Φ *(columns)*

k_0	k_{10}	k_4	k_{12}	a'_0	b'_0	c_0	c'_0	b_0	a_0
K_0	K_{10}	K_4	K_{12}	A'_2	B'_5	C_6	C'_6	B_5	A_2
K_3	K_0	K_7	K_2	A'_5	B'_8	C_9	C'_9	B_8	A_5
K_9	K_6	K_0	K_8	A'_{11}	B'_1	C_2	C'_2	B_1	A_{11}
K_1	K_{11}	K_5	K_0	A'_3	B'_6	C_7	C'_7	B_6	A_3
A_0	A_{10}	A_4	A_{12}	K_0	A'_9	B'_{11}	C_5	C'_{11}	B_9
B_0	B_{10}	B_4	B_{12}	A_6	K_0	A'_8	B'_3	C_3	C'_8
C'_0	C'_{10}	C'_4	C'_{12}	B_7	A_7	K_0	A'_1	B'_2	C_1
C_0	C_{10}	C_4	C_{12}	C'_1	B_2	A_1	K_0	A'_7	B'_7
B'_0	B'_{10}	B'_4	B'_{12}	C_8	C'_3	B_3	A_8	K_0	A'_6
A'_0	A'_{10}	A'_4	A'_{12}	B'_9	C_{11}	C'_5	B_{11}	A_9	K_0

Exercise 3.6.3 Rearrange Table 3.5.2 with columns in the order $(ABC'CB'A')$ and rows in the order $(abc'cb'a')$ and number the rows and columns 1, 2, ..., 6. (This table is printed later as Table 3.8.1.) Verify that, if $\{x\}$ is the entry or set of entries in (column r, row s), then $\{3x, \bmod 13\}$ is the entry or set of entries in (column $3r$, row $3s$, mod 7).

Under this transformation

$$K_r \to K_{3r} \quad \text{and} \quad A_r \to C'_{3r} \to B_{9r} \to A'_r \to C_{3r} \to B'_{9r} \to A_r.$$

Denote the transformation by $\mathcal{T}$: then $\mathcal{T}^3$ ($A_r \to A'_r$, etc.) is the collineation induced by the interchange of complex conjugates, and $\mathcal{T}^4 (K_r \to K_{3r}, A_r \to C_{3r} \to B_{9r} \to A_r, A'_r \to C'_{3r} \to B'_{9r} \to A'_r)$ is, like $\mathcal{T}$, an extension to Φ of the collineation $K_r \to K_{3r}$ in Δ_0.

Finally let us rewrite Table 3.5.2, replacing the symbols $A_r, B_r, \ldots$ by the position numbers of the points in the Singer cycle generated by $\mathbf{W}$ (Table 3.6.4).

Table 3.6.6 *Collinear sets of points* $W^t(1, 0, 0)$, $t = 7r + s$ *(collinear sets by rows)*

	s: 1	2	3	4	5	6
1	22, 29 50, 64	65	73	67	54	34
2	57	16, 79 86, 30	31	39	33	20
3	71	37	66, 87 59, 10	11	19	13
4	78	58	24	53, 74 46, 88	89	6
5	85	72	52	18	68, 40 47, 82	83
6	8	2	80	60	26	48, 55 76, 90

(i) The set of 54 numbers in the table is the complement in the set $\{0, \ldots, 90\}$ of the set

$$\{7r, s, s+42, s+35, s+14 \colon r = 0, \ldots, 12, s = 1, 3, 9, 27, 81, 61\}.$$

(ii) The set of 12 differences between elements in the same columns in any two rows is $\{7, 14, \ldots, 84 \colon \bmod 91\}$.

(iii) The set of differences obtained by subtracting elements in column s' from elements in the same rows in column s is $\{7r+s-s': r = 1, \ldots, 12\}$.

(iv) If $\{x\}$ is the number or set of numbers in (row r, column s) then $\{3x, \bmod 91\}$ is the number or set of numbers in (row $3r$, column $3s$, mod 7).

3.7 Conics in Φ

Let $\mathbf{x}^T\mathbf{C}\mathbf{x}$ be any non-singular quadratic form over $\mathscr{F}$; in Φ we can define a polarity $\mathscr{C}$ in relation to this form in which, for every point $\mathbf{p}$, $\mathscr{C}\mathbf{p}$ is $\mathbf{p}^T\mathbf{C}\mathbf{x} = 0$. By selecting as triangle of reference a triangle which is self-polar in this polarity we can reduce the form to $\omega^t(\omega^r x^2 + \omega^s y^2 + z^2)$. For every pair of values $\{r, s\}$ we can readily identify at least one point for which

$$\omega^r x^2 + \omega^s y^2 + z^2 = 0.$$

For example, if $\{r, s\} = \{3, 2\}$, we have $\omega^7 + \omega^2 + 1 = 0$, so that for these values the equation is satisfied by $(\omega^2, 1, 1)$. There is therefore, by Theorem 2.6.6, a set of ten points which lie on their polar lines. So that:

THEOREM 3.7.1 *If* $\mathbf{C}$ *is non-singular, the set of points*

$$\{\mathbf{x}: \mathbf{x}^T\mathbf{C}\mathbf{x} = 0\}$$

in Φ *always contains ten points.*

No three of the ten points are collinear (since if three points of a line satisfy $\mathbf{x}^T\mathbf{C}\mathbf{x} = 0$, then every point of the line does so, and then $\mathbf{C}$ is singular), so that any four of them form a regular quadrangle and may be used as a basis for coordinates. Thus for any given conic in Φ the defining matrix may be chosen to be either of the matrices:

$\mathbf{C} = \mathbf{1}$, conic $x^2+y^2+z^2 = 0$, points $(\pm 1, \pm 1, 1)$, $(\pm\omega^2, 1, 0)$, $(\pm\omega^2, 0, 1)$, $(0, \pm\omega^2, 1)$.

$\mathbf{C} = \begin{bmatrix} 0 & 0 & 1 \\ 0 & 1 & 0 \\ 1 & 0 & 0 \end{bmatrix}$, conic $y^2 - xz = 0$, points $(1, 0, 0), (0, 0, 1)$, $(\omega^{2r}, \omega^r, 1)$ $(r = 0, \ldots, 7)$.

We can define interior and exterior points as in Definition 2.6.4 and then from Theorem 2.6.7 and Exercise 2.6.6 we find, since -1 is a square in $\mathscr{F}$:

THEOREM 3.7.2 *The sets* Γ (*on the conic*), Γ_e (*exterior*), Γ_i (*interior*) *are*

Γ: $\{\mathbf{x}: \mathbf{x}^T\mathbf{C}\mathbf{x} = 0\}$, 10 *points, each on its polar.*

Γ_e: $\{\mathbf{x}$: $(\det \mathbf{C})$ $\mathbf{x}^T\mathbf{C}\mathbf{x} \in \{\pm 1, \pm \omega^2\}\}$, 45 *points each such that its polar contains two points of* Γ.

Γ_i: $\{\mathbf{x}$: $(\det \mathbf{C})\,\mathbf{x}^T\mathbf{C}\mathbf{x} \in \{\pm \omega, \pm \omega^3\}\}$, 36 *points each such that its polar contains no point of* Γ.

EXERCISE 3.7.1 A non-singular conic in a plane of order n contains $n+1$ points. Therefore a non-singular conic in Δ, the plane of order 3, contains only four points; that is, the non-singular conics in Δ are the regular quadrangles. Prove the following properties of conics in Δ:

(i) The conic $x^2+y^2+z^2 = 0$ is, in the notation of Exercise 3.2.4, $\{K_5, K_6, K_7, K_{11}\}$; the tangents are $\{k_9, k_3, k_7, k_8\}$.

(ii) The interior points of a non-singular conic in Δ are the three intersections of pairs of opposite sides of the quadrangle.

(iii) A singular conic in Δ consists either of two lines, a single line, or a single point.

(iv) The total number of distinct conics represented by the vanishing of quadratic forms $\Sigma c_{ij} x_i x_j$ over GF(3) is

$$\tfrac{1}{2}(3^6-1) = 364.$$

How many of these are (*a*) regular quadrangles, (*b*) line-pairs, (*c*) single lines, (*d*) single points?

EXERCISE 3.7.2 (i) Prove that the number of ordered sets of 5 points in Φ, no three of which are collinear, is $91.90.81.64.42$.

(ii) If a quadratic form $\phi(x,y,z)$ vanishes at such a set of points, then the set $\{\mathbf{x}: \phi(x,y,z) = 0\}$ is a non-singular conic, consisting of 10 points, no three collinear. Thence prove that the number of non-singular conics in Φ is $91.9^2.8$.

(iii) Check this number by counting (*a*) the number of distinct equations $\phi = 0$, (*b*) the numbers of line-pairs and single lines, (*c*) the number of distinct forms ϕ which vanish at only one point.

EXERCISE 3.7.3 From Table 3.5.2 (or, better, from its re-arrangement as Table 3.8.1) show that

$$K_r \to k_{-r}, \quad A_r \to a'_{-r}, \quad A'_r \to a_{-r}, \quad B_r \to b'_{-r}, \quad \text{etc.},$$

$r \in \{0, 1, \ldots, 12\}$, is a correlation. Prove that the points which lie on their correlative lines are $K_0, K_7, K_8, K_{11}, A_3, A'_3, B_1, B'_1, C_9, C'_9$, and that the correlation is the polarity with regard to the conic $y^2+z^2-xz = 0$.

EXERCISE 3.7.4 Prove that a non-singular conic may have 4, 3, 2, 1 or 0 points common with a subplane of order 3 in Φ. (Select a coordinate-system such that the conic is the set of points $\{(\kappa^2, \kappa, 1): \kappa \in \mathscr{F}\} \cup \{(1, 0, 0)\}$. The subplane is the set of points $\{(\mathbf{M}+\epsilon\mathbf{N})\,\mathbf{x}: \mathbf{x} \text{ is real}\}$, for some real $\mathbf{M}, \mathbf{N}$. Now choose suitable pairs $\mathbf{M}, \mathbf{N}$ to show that all five possibilities occur.)

3.8 Hermitian sets in Φ

In Φ there are polarities which are not associated with quadratic forms, namely those derived from 'Hermitian forms' over $\mathscr{F}$.

If we combine a projective correlation $\mathbf{p} \to \mathbf{p}^T\mathbf{M}\mathbf{x} = 0$ with the non-projective collineation $\mathscr{J}$: $\mathbf{x} \to \mathbf{x}^*$ we obtain a non-projective correlation

$$\mathbf{p} \to \mathbf{p}^{*T}\mathbf{M}\mathbf{x} = 0.$$

The set of points incident with their correlative lines in this correlation is given by $\mathbf{x}^{*T}\mathbf{M}\mathbf{x} = 0$. We may split $\mathbf{M}$ into matrices corresponding to the matrices $\mathbf{C}$ and $\mathbf{S}$ in Theorem 2.7.4 in the following way:

$$\mathbf{M} = \mathbf{H}+\epsilon\mathbf{L}, \quad \text{where} \quad \mathbf{H} = \tfrac{1}{2}(\mathbf{M}+\mathbf{M}^{*T}), \quad \epsilon\mathbf{L} = \tfrac{1}{2}(\mathbf{M}-\mathbf{M}^{*T}).$$

Then

$$\mathbf{H}^{*T} = \mathbf{H}, \quad (\epsilon\mathbf{L})^{*T} = -\epsilon\mathbf{L}, \quad \text{that is} \quad \mathbf{L}^{*T} = \mathbf{L}.$$

DEFINITION 3.8.1 A *Hermitian matrix* is a matrix $\mathbf{H}$ such that $\mathbf{H}^{*T} = \mathbf{H}$.

The two matrices $\mathbf{H}$ and $\mathbf{L}$ above are Hermitian.

THEOREM 3.8.1 *In the correlation* $\mathbf{p} \rightarrow \mathbf{p}^{*T}\mathbf{M}\mathbf{x} = 0$, *the set of points which lie on their correlative lines is the intersection of the two point-sets given by* $\mathbf{x}^{*T}\mathbf{H}\mathbf{x} = 0$ *and* $\mathbf{x}^{*T}\mathbf{L}\mathbf{x} = 0$, *where* $\mathbf{M} = \mathbf{H} + \epsilon\mathbf{L}$ *and* $\mathbf{H}$ *and* $\mathbf{L}$ *are Hermitian.*

Let us consider the equation $\mathbf{x}^{*T}(\mathbf{H}+\epsilon\mathbf{L})\,\mathbf{x} = 0$. Now

$$(\mathbf{x}^{*T}\mathbf{H}\mathbf{x})^* = \mathbf{x}^T\mathbf{H}^*\mathbf{x}^* = (\mathbf{x}^T\mathbf{H}^*\mathbf{x}^*)^T = \mathbf{x}^{*T}\mathbf{H}^{*T}\mathbf{x} = \mathbf{x}^{*T}\mathbf{H}\mathbf{x}$$

so that, for all vectors $\mathbf{x}$, $\mathbf{x}^{*T}\mathbf{H}\mathbf{x}$ and likewise $\mathbf{x}^{*T}\mathbf{L}\mathbf{x}$ are real, and therefore $\mathbf{x}^{*T}(\mathbf{H}+\epsilon\mathbf{L})\,\mathbf{x} = 0 \Leftrightarrow \mathbf{x}^{*T}\mathbf{H}\mathbf{x} = 0$ and $\mathbf{x}^{*T}\mathbf{L}\mathbf{x} = 0$. The greatest interest attaches therefore to the type of point-set satisfying the single condition $\mathbf{x}^{*T}\mathbf{H}\mathbf{x} = 0$.

DEFINITION 3.8.2 A *non-singular Hermitian set* $\mathscr{H}$ is a point-set $\{\mathbf{x}: \mathbf{x}^{*T}\mathbf{H}\mathbf{x} = 0\}$ with $\mathbf{H}$ Hermitian and non-singular.

$\mathscr{H}$ consists of the points which lie on their correlative lines in the non-projective correlation $\mathbf{p} \rightarrow \mathbf{p}^{*T}\mathbf{H}\mathbf{x} = 0$. The Hermitian property of $\mathbf{H}$ implies in fact that this correlation is a polarity: for

$$\mathbf{p} \in \mathbf{q}^{*T}\mathbf{H}\mathbf{x} = 0 \Rightarrow \mathbf{q}^{*T}\mathbf{H}\mathbf{p} = 0 \Rightarrow \mathbf{q}^T\mathbf{H}^*\mathbf{p}^* = 0 \Rightarrow \mathbf{p}^{*T}\mathbf{H}^{*T}\mathbf{q} = 0$$
$$\Rightarrow \mathbf{p}^{*T}\mathbf{H}\mathbf{q} = 0 \Rightarrow \mathbf{q} \in \mathbf{p}^{*T}\mathbf{H}\mathbf{x} = 0.$$

Now let $\mathbf{u}^T\mathbf{x} = 0$ be the polar of $\mathbf{p}$, then $\mathbf{p}^{*T}\mathbf{H} = \rho^*\mathbf{u}^T$; that is $\mathbf{p} = \rho(\mathbf{H}^{*T})^{-1}\,\mathbf{u}^* = \rho\mathbf{H}^{-1}\mathbf{u}^*$; each line of the dual Hermitian set $\{\mathbf{u}^T\mathbf{x} = 0: \mathbf{u}^{*T}\mathbf{H}^{*-1}\mathbf{u} = 0\}$ is the polar of the point $\mathbf{H}^{-1}\mathbf{u}^*$ of the Hermitian set $\{\mathbf{x}: \mathbf{x}^{*T}\mathbf{H}\mathbf{x} = 0\}$. Thus a non-singular Hermitian matrix determines a polarity which separates the points and lines of the plane into two sets; namely,

Absolute point (a point of the Hermitian set): a point which lies on its polar line,

Exterior point: a point which does not lie on its polar line,

Tangent line (a line of the dual Hermitian set): a line which contains its pole,

Secant line: a line which does not contain its pole.

We are to prove:

Theorem 3.8.2 *Each tangent of $\mathscr{H}$ contains only one absolute point, namely its pole; each secant contains four absolute points. Dually each absolute point lies on a single tangent line, and each exterior point lies on four tangents.*

In relation to a form $\mathbf{x}^{*T}\mathbf{H}\mathbf{x}$, where

$$\mathbf{H} = \begin{bmatrix} a & \nu & \mu^* \\ \nu^* & b & \lambda \\ \mu & \lambda^* & c \end{bmatrix}$$

is an unrestricted non-singular Hermitian matrix, any given line may be assumed to have the equation $x = 0$. This line intersects the Hermitian set $\mathbf{x}^{*T}\mathbf{H}\mathbf{x} = 0$ in the points $(0, y, z)$ given by

$$byy^* + \lambda y^* z + \lambda^* y z^* + czz^* = 0. \qquad \text{(i)}$$

Assume first that $b \neq 0$, and adjust the coefficients so that $b = 1$. Since $(0, 1, 0)$ is not a solution of the equation, we may take the solutions in the form $(0, y, 1)$. Write $y = y_1 + \epsilon y_2$ and $\lambda = l_1 + \epsilon l_2$, then the relation becomes:

$$(y+\lambda)(y^*+\lambda^*) + c - \lambda\lambda^*$$
$$\equiv (y_1+l_1)^2 + (y_2+l_2)^2 + c - l_1^2 - l_2^2 = 0. \qquad \text{(ii)}$$

The solutions are:

when $\quad c - l_1^2 - l_2^2 = 1$: $y_1 + l_1 = \pm 1, \quad y_2 + l_2 = \pm 1, \qquad$ (iii$a$)

$c - l_1^2 - l_2^2 = -1$: $\left.\begin{array}{l} y_1 + l_1 = \pm 1, \quad y_2 + l_2 = 0, \\ y_1 + l_1 = 0, \quad y_2 + l_2 = \pm 1, \end{array}\right\} \qquad$ (iii$b$)

$c - l_1^2 - l_2^2 = 0$: $y_1 + l_1 = 0, \quad y_2 + l_2 = 0. \qquad$ (iii$c$)

Thus, when $b \neq 0$, there are on $x = 0$ either four (distinct) points or a single point of $\mathscr{H}$. If $b = 0$ we may change the co-ordinate system leaving $x = 0$ fixed, say by the substitution $(y, z) \to (\rho y + \sigma z, \rho' y + \sigma' z)$, in such a way that equation (i) is replaced by a relation in which the coefficient of yy^* is not zero.

Consider now the case in which $x = 0$ contains only one absolute point, namely, from equation (iii$c$),

$$\mathbf{H} = \begin{bmatrix} a & \nu & \mu^* \\ \nu^* & 1 & \lambda \\ \mu & \lambda^* & \lambda\lambda^* \end{bmatrix}.$$

The single point is given by $y = -l_1 - \epsilon l_2 = -\lambda$, that is, the point is $(0, -\lambda, 1)$, and its polar line is $[0, -\lambda^*, 1]\mathbf{Hx} = 0$, namely $x = 0$. This completes the proof of Theorem 3.8.2, which in turn enables us to prove:

THEOREM 3.8.3 *$\mathscr{H}$ has 28 points* and 28 *tangents,* 63 *exterior points and* 63 *secants.*

Through any absolute point T there is a single tangent, so that the other nine lines through T are secants, each of which contains three absolute points other than T, 27 additional absolute points altogether. Since every point of Φ and in particular every absolute point lies on the tangent through T or one of the secants through T, the total number of absolute points is 28; the rest of the theorem follows immediately. Each tangent contains one absolute point and nine exterior points; each secant contains four absolute points and six exterior points.

EXERCISE 3.8.1 P is an exterior point relative to $\mathscr{H}$. Through P there are six secants containing altogether 24 absolute points. Prove that the remaining four absolute points lie on the polar line of P.

EXERCISE 3.8.2 In Exercise 3.8.1 we have found a set of seven secants which contains all 28 absolute points, but six of the secants are concurrent. Is it possible to find a set of seven secants with 21 distinct points of intersection which contains all 28 absolute points? (Difficult: the answer is no.)

EXERCISE 3.8.3 (i) Prove that it is possible to find a self-polar triangle for the Hermitian system, namely a set of three non-collinear points the join of each pair of which is the polar of the third (cf. Theorem 2.6.5). (ii) Prove that the number of self-polar triangles is 63.

Take as triangle of reference a self-polar triangle; since $x_i = 0$ is the polar of $\mathbf{e}_i$, the Hermitian form $\mathbf{x}^{*T}\mathbf{Hx}$ is such that

$$h_{ij} = 0 \quad \text{if} \quad i \neq j, \quad \text{and} \quad h_{ii} \neq 0;$$

that is, it reduces to $\Sigma(\pm x_i^* x_i) = 0$. For each $-$ve sign replace x_i by ωx_i so that $x_i^* x_i$ is replaced by $\omega^3 x_i^* . \omega x_i = -x_i^* x_i$, and the equation of the Hermitian set $\mathscr{H}$ becomes

$$\mathbf{x}^{*T}\mathbf{x} \equiv x^*x + y^*y + z^*z = 0.$$

Since the three terms in the sum are real we must have either:

$$\text{(i)}\quad xx^* = yy^* = zz^* = 1,$$

or one of the six sets of conditions symmetrical with

$$\text{(ii)}\quad xx^* = -1, \quad yy^* = 1, \quad zz^* = 0.$$

For (i) we have: $xx^* = 1 \Rightarrow x \in \{\pm 1, \pm \omega^2\}$. This condition therefore provides 16 points $(\pm 1, \pm 1, 1)$, $(\pm \omega^2, \pm 1, 1)$, $(\pm 1, \pm \omega^2, 1)$, $(\pm \omega^2, \pm \omega^2, 1)$. From (ii): $xx^* = -1$, $yy^* = 1$, $zz^* = 0$ gives the four points $(\pm \omega, 1, 0)$, $(\pm \omega^3, 1, 0)$. But the six sets of four points coincide in pairs.

To find a parametric form for $\mathscr{H}$ we construct the equation which is the analogue of the conic-equation $y^2 + 2xz = 0$, namely $yy^* + xz^* + zx^* = 0$. Denote the corresponding Hermitian set by Γ_H. The coordinate-triangle for Γ_H consists of an arbitrary exterior point, taken as $(0, 1, 0)$, with two of the four absolute points on its polar taken as $(1, 0, 0)$ and $(0, 0, 1)$, and the coefficients are fixed by taking $(1, 1, 1)$ to be an absolute point. The points of Γ_H which lie on $y = \tau z$ are given by

$$\tau\tau^* zz^* = -xz^* - zx^*$$

or
$$\tau\tau^* = -x/z - (x/z)^*. \qquad \text{(i)}$$

Write
$$\tau = t + \epsilon t', \quad x/z = v + \epsilon u,$$

then equation (i) requires

$$\tau\tau^* = t^2 + t'^2 = -2v = v,$$

but imposes no restriction on u. Thus:

THEOREM 3.8.4 *The Hermitian set*

$$\Gamma_H = \{\mathbf{x} : yy^* + xz^* + zx^* = 0\}$$

consists of $(1, 0, 0)$ *together with the* 27 *points*

$$\{(t^2 + t'^2 + \epsilon u, t + \epsilon t', 1) : t, t', u \in \{0, \pm 1\}\}.$$

Finally we derive the Hermitian set in a completely different way by observing properties of the table of incidences constructed in Exercise 3.6.3, as shown by Table 3.8.1.

Table 3.8.1

$k_r = \{K_r, K_{r+1}, K_{r+3}, K_{r+9}, A_r, B_r, C_r, A'_r, B'_r, C'_r\}.$

$U_r \in v_0$	A_r	B_r	C'_r	C_r	B'_r	A'_r
a_0	2, 3, 5, 11	9	8	1	7	6
b_0	9	5, 6, 8, 1	11	3	2	7
c'_0	8	11	6, 7, 9, 2	5	3	1
c_0	1	3	5	6, 7, 9, 2	11	8
b'_0	7	2	3	11	5, 6, 8, 1	9
a'_0	6	7	1	8	9	2, 3, 5, 11

$$U_r \in v_0 \Rightarrow U_{r+a} \in v_a.$$

Since
$$U_s \in v_t \Rightarrow U_{s-(s+t)} \in v_{t-(s+t)}$$
$$\Rightarrow U_{-t} \in v_{-s},$$
there is a correlation of Φ given by

$$K_r \to k_{-r}, \quad U_r \to u_{-r}, \quad U \in \{A, B, C, A', B', C'\},$$
$$u \in \{a, b, c, a', b', c'\}.$$

Further, since $U_r \in u_0 \Rightarrow U_{\frac{1}{2}r} \in u_{-\frac{1}{2}r}$, the absolute points are given by the points whose indices are the halves of those in the leading diagonal in the table above, namely,

K_r: $r = 0, 7, 8, 11,$ $\qquad$ A_r, A'_r: $r = 1, 9, 8, 12,$

B_r, B'_r: $r = 9, 3, 7, 4,$ $\qquad$ C_r, C'_r: $r = 3, 1, 11, 10.$

We have to verify that this is a Hermitian set. First, the four real points are identical with those on the conic constructed in Exercise 3.7.3, and the equation of this conic may be written as $y^2+z^2+2xz = 0$. We need verify therefore only that the other points of the set above satisfy the condition

$$yy^*+zz^*+xz^*+zx^* = 0.$$

Thus

A_1 is $(0, \omega, 1)$, a_{12} is $x+\omega^3 y+z = 0$,

A_8 is $(-\omega^2, \omega^3, 1)$, a_5 is $x+\omega y+\omega^3 z = 0$, etc.

In each case the point belongs to the Hermitian set and the line is the corresponding tangent. For tactical investigations the naming of points and lines as in Table 3.8.1 is probably the easiest to use.

EXERCISE 3.8.4 The transformation

$$A_r \to C'_{3r} \to B_{9r} \to A'_r \to C_{3r} \to B'_{9r} \to A_r, \quad K_r \to K_{3r} \to K_{9r} \to K_r,$$

splits the 63 secants into nine sets of 6 and three sets of 3. For example:

Set of 3			Set of 6					
K_0	K_0	K_0	A_1	C'_3	B_9	A'_1	C_3	B'_9
K_7	K_8	K_{11}	A_8	C'_{11}	B_7	A'_8	C_{11}	B'_7
B_4	A'_{12}	C_{10}	A'_9	C_1	B'_3	A_9	C'_1	B_3
B'_4	A_{12}	C'_{10}	C'_{11}	B_7	A'_8	C_{11}	B'_7	A_8
k_4	k_{12}	k_{10}	a_3	c'_9	b_1	a'_3	c_9	b'_1

Construct a table showing one member of each set and investigate the figures formed by the sets of points and lines.

EXERCISE 3.8.5 In Exercise 3.8.1 take $P = K_0$ and write down the seven collinear sets of four absolute points.

EXERCISE 3.8.6 The notation and analysis associated with Theorems 3.8.1 to 3.8.4 can be applied directly to Π_{q^2}, the plane over GF (q^2), $q = p^n$, p an odd prime. Prove that: (i) a non-singular Hermitian matrix in Π_{q^2} separates the points and lines of the plane into absolute points, exterior points, tangents and secants, (ii) a tangent contains exactly one absolute point and a secant contains exactly $q+1$ absolute points, (iii) the number of absolute points is q^3+1, (iv) through any exterior point P there pass q^2-q secants containing altogether q^3-q absolute points, and the remaining $q+1$ absolute points lie on the polar line of P.

PART III

MINIQUATERNION PLANES

All the planes described in Part II are Desarguesian. As an introduction to non-Desarguesian planes, we describe in this Part three planes of order 9 which we call 'miniquaternion planes' because they can be coordinatized (each in a different, but not unduly complicated way) by the miniquaternion near-field. These planes were discovered by O. Veblen and J. M. Wedderburn (1907), shortly after the construction by L. E. Dickson (1905) of a large class of near-fields which includes the miniquaternions. One of the planes is a translation plane: we call this plane Ω, and its dual plane Ω^D. The third plane, to be called Ψ, is neither a translation plane nor the dual of a translation plane. Ψ is self-dual; that is, $\Psi^D = \Psi$.

Ω and Ω^D may be coordinatized by miniquaternions in very similar (and natural) ways. G. Zappa (1957) and T. G. Ostrom (1964) discovered two different methods of coordinatizing Ψ. Interest in Ψ has been revived by D. R. Hughes (1957), who considered an infinite class of planes which includes Ψ (the 'Hughes planes'). We shall give a definition of Ψ which enables us to derive the Ostrom coordinatization almost immediately.

The coordinatizations of the three miniquaternion planes will be based on analogues in miniquaternion algebra of three different ways of writing the equation of a line in a field-plane.

To avoid reference back we repeat here the main features of the miniquaternion algebra that were established in §1.3:

$$\mathscr{Q} = \{0, \pm 1, \pm\alpha, \pm\beta, \pm\gamma\},$$

$$\mathscr{D} = \{0, \pm 1,\}, \quad \mathscr{Q}^* = \{\pm\alpha, \pm\beta, \pm\gamma\}.$$

The algebra $\mathscr{Q}$ is defined by an additive group $\mathscr{Q}_A$, and a multiplicative group $\mathscr{Q}_M$ the structures of which induce the rules of distribution.

$$\mathscr{Q}_A\colon \xi+\xi+\xi = 0 \quad \text{for all } \xi,$$

$$\beta = \alpha - 1, \quad \gamma = \alpha + 1,$$

whence $\alpha-\beta = \beta-\gamma = \gamma-\alpha = 1, \quad \alpha+\beta+\gamma = 0,$

$$\mathscr{Q}_M: \alpha^2 = \beta^2 = \gamma^2 = \alpha\beta\gamma = -1,$$

whence

$$\beta\gamma = -\gamma\beta = \alpha, \quad \gamma\alpha = -\alpha\gamma = \beta, \quad \alpha\beta = -\beta\alpha = \gamma.$$

Distributive rules:

$$(\rho+\sigma)\tau = \rho\tau+\sigma\tau \quad \text{for all} \quad \rho, \sigma, \tau$$

but in general $\tau(\rho+\sigma) \neq \tau\rho+\tau\sigma.$

We shall refer to $\mathscr{D} = \{0, \pm 1\}$ as the set of real elements and $\mathscr{Q}^* = \{\pm\alpha, \pm\beta, \pm\gamma\}$ as the set of complex elements, and in general we shall use Italic letters $a, b, r, \ldots$ for real elements and Greek letters (other than α, β, γ) for elements of $\mathscr{Q} = \mathscr{D}+\mathscr{Q}^*$, except that we shall use (x, y) or (x, y, z) as coordinate-vectors over $\mathscr{Q}$, and sometimes shall restrict ρ, σ, τ to being members of $\mathscr{Q}^*$.

We find right at the outset that we cannot use homogeneous coordinates without some modification, for suppose that the coordinate-vectors $[x, y, z]$, $[\lambda, \mu, \nu]$ represent respectively a point and a line, and that the conditions of incidence are written either as

$$[\lambda, \mu, \nu]\begin{bmatrix}x\\y\\z\end{bmatrix} = 0 \quad \text{or} \quad [x, y, z]\begin{bmatrix}\lambda\\\mu\\\nu\end{bmatrix} = 0;$$

that is, as

$$\lambda x+\mu y+\nu z = 0 \tag{i}$$

or

$$x\lambda+y\mu+z\nu = 0. \tag{ii}$$

Consider case (i): if the coordinates are homogeneous, then for given (λ, μ, ν) the same set of points must satisfy the condition

$$(\kappa\lambda)x+(\kappa\mu)y+(\kappa\nu)z = 0 \quad (\kappa \neq 0). \tag{i$'$}$$

But since miniquaternions are not left-distributive, so that, if $\kappa \neq \pm 1$, $\kappa(\lambda x+\mu y+\nu z) \neq (\kappa\lambda x+\kappa\mu y+\kappa\nu z)$, the sets of points represented by (i) and (i′) are different.

Likewise in case (ii) the non-distributive property prevents the use of homogeneous coordinates for points; we can, however, find a valid form of equation in Cartesian-type coordinates. We isolate a line, say $z = 0$, which is to behave differently from other lines and replace the projective plane by the affine plane, which is the projective plane with this line obliterated. Each set of lines in the projective plane through a point of $z = 0$ is replaced by a set of parallel lines in the affine plane.

In relation to such a system in a field-plane we may take the equation of a line to be in the classical Cartesian form

$$y = x\mu + \kappa, \tag{iii}$$

while lines to which this form is not applicable, namely those parallel to $x = 0$, have equations

$$x = \lambda.$$

The ideal line is of course not included (the equations $y = x\mu + \kappa$, $y = x\mu + \kappa'$, with $\kappa \neq \kappa'$, are inconsistent) but can be referred to as $[0, 0, 1]$ or $z = 0$. Since $x\mu + \kappa \equiv (x + \kappa\mu^{-1})\mu$, $\mu \neq 0$, we may write the two forms of equations together in quasi-homogeneous form as

$$[x, y, 1](\mu, e, \kappa) = 0 \quad \text{where} \quad e = -1 \quad \text{or} \quad 0.$$

The line is specified by:

the point $(0, \kappa)$ in which it meets $x = 0$, and

its 'slope' (μ), which is in effect the point on the ideal line with quasi-homogeneous coordinates $(1, \mu, 0)$.

The plane Ω is defined as the miniquaternion analogue of the field-plane with these conditions of incidence of point and line.

We saw that in the field-plane Φ every complex line contains exactly one real point. Let us therefore replace condition (iii) for incidence in Φ by a condition in terms of the slope of the line and the real point on the line. That is, instead of the equation (iii), we experiment with the equation

$$y - s = (x - r)\mu \quad (r, s \in \mathscr{D}) \tag{iv}$$

together with $x = \lambda$, $z = 0$ as before. To find the relation between conditions (iii) and (iv), write (iv) as

$$y = x\mu + s - r\mu,$$

then the same line is represented by the two forms of equation if and only if $\kappa = s - r\mu$. Provided μ is complex, this is a 1–1 correspondence between complex elements κ and ordered pairs (r, s), where $s \in \{0, \pm 1\}, r \in \{\pm 1\}$.

If, however, μ is real, say $\mu = m$, and κ is complex, the equation $y = xm + \kappa$ cannot be written in form (iv) since this would involve the contradictory condition $\kappa = s - rm$, so that, to obtain an equation for every line in Φ, we could use the set of equations

$$y - s = (x - r)\mu \quad (\mu \in \mathscr{F} - \mathscr{D}), \tag{iv}$$

$$y = xm + \kappa \quad (\kappa \in \mathscr{F}), \tag{iv$'$}$$

together with $x = \lambda$ and $z = 0$.

The replacement of field elements by miniquaternion elements does not affect the analysis above, since the crucial relation $\kappa = s - r\mu$ does not involve the products of complex elements. That is, the forms of condition (iv), (iv′) apply also to the miniquaternion plane Ω.

Let us now turn to the two corresponding types of equations in which the slope operates as a left-multiplier instead of a right-multiplier, taking first

$$y = \mu x + \kappa \tag{v}$$

(and $x = \lambda$, $z = 0$). As an equation in miniquaternions, (v) may be replaced by

$$(y\rho) = \mu(x\rho) + (\kappa\rho);$$

that is, coordinates of points are effectively homogeneous, and no particular line can be singled out as the ideal line. On the other hand, condition (v) cannot be replaced by $y = \sigma^{-1}[(\sigma\mu)x + \sigma\kappa]$, so that μ and κ are absolute, and the point $(0, 1, 0)$ is special. As we might expect, and as we shall prove in §4.7, the plane, in which the condition of incidence of the point (x, y, z) with the line $[\mu, e, \kappa]$, $e = 0$ or -1, is $[\mu, e, \kappa](x, y, z) = 0$, turns out to be the dual Ω^D of the plane Ω.

Finally let us consider the set of equations

$$y - s = \mu(x - r) \quad (\mu \in \mathscr{Q}^*), \tag{vi}$$

$$y = mx + \kappa \quad (\kappa \in \mathscr{Q}), \tag{vi$'$}$$

together with $x = \lambda$ and $z = 0$. This is the most awkward of the sets of conditions to handle; in Chapter 5 we shall prove that there is a plane Ψ for which these are the conditions of incidence, and that this plane is self-dual (and therefore different from Ω and Ω^D).

We may tabulate the incidence conditions in the miniquaternion planes in the following way:

Ω^D	Ω	Ψ
Line with complex slope μ through $(0, \kappa, 1)$		
$y = \mu x + \kappa$	$y = x\mu + \kappa$	—
	Line with complex slope μ through the real point $(r, s, 1)$	
—	$y - s = (x - r)\,\mu$	$y - s = \mu(x - r)$
Line with real slope m through $(0, \kappa, 1)$		
$y = mx + \kappa$	$y = mx + \kappa$	$y = mx + \kappa$
Other lines		
$x = \kappa,\ z = 0$	$x = \kappa,\ z = 0$	$x = \kappa,\ z = 0$

CHAPTER 4

THE PLANES Ω AND Ω^D

4.1 The construction of the plane Ω

Let us consider first the set of 81 points $\{(x,y): x, y \in \mathscr{Q}\}$ and the 90 lines defined to be the subsets of this set of points which satisfy one of the two conditions

$$y = x\mu + \kappa \quad (\mu, \kappa \in \mathscr{Q}) \quad (81),$$

$$x = \lambda \qquad (\lambda \in \mathscr{Q}) \quad (9).$$

Later these points and lines are to be designated the 'proper points' and 'proper lines' of the plane Ω. We prove:

THEOREM 4.1.1 (i) *Joining two distinct points there is exactly one line.* (ii) *Given a line l and a point A not on l, there is exactly one line through A which does not meet l.*

(i) Let $P_1 = (x_1, y_1)$ and $P_2 = (x_2, y_2)$ be two distinct points. If $x_1 = x_2$ the unique line joining P_1 and P_2 is $x = x_1$. (There can be no line $y = x\mu + \kappa$ through both P_1 and P_2 since $y_1 \neq y_2$ whereas $x_1 = x_2$.) If $x_1 \neq x_2$ then P_1 and P_2 both lie on a line $y = x\mu + \kappa$ if and only if

$$y_1 = x_1\mu + \kappa, \quad y_2 = x_2\mu + \kappa.$$

Remembering that multiplication is neither commutative nor left-distributive over addition, we reduce these equations to the equivalent pairs

$$y_1 - y_2 = x_1\mu - x_2\mu, \quad y_2 = x_2\mu + \kappa,$$

or
$$y_1 - y_2 = (x_1 - x_2)\,\mu, \quad y_2 = x_2\mu + \kappa,$$

or $\quad \mu = (x_1 - x_2)^{-1}(y_1 - y_2), \quad \kappa = y_2 - x_2(x_1 - x_2)^{-1}(y_1 - y_2).$

The last pair of equations shows explicitly that μ and κ are uniquely determined by P_1 and P_2. (There can be no line $x = \lambda$ through P_1 and P_2 since $x_1 \neq x_2$.)

(ii) Let $A = (a, b)$. If l is $y = x\mu + \kappa$ then $y = x\mu + (b - a\mu)$ is the required line; while if l is $x = \lambda$ then $x = a$ is the required line. (Since $A \notin l$, $b - a\mu \neq \kappa$ in the first case, and $\lambda \neq a$ in the second case.) Every line $x = \lambda_1$ meets every line $y = x\mu_1 + \kappa_1$, and the lines $y = x\mu_1 + \kappa_1$ and $y = x\mu_2 + \kappa_2$ have a common point if $\mu_1 \neq \mu_2$, by Theorem 1.5.3. Hence the uniqueness.

The families of 'parallel' lines in this geometrical system are: for each $\mu \in \mathcal{Q}$, the nine lines $y = x\mu + \kappa$ (κ varying over $\mathcal{Q}$); the nine lines $x = \lambda$ (λ varying over $\mathcal{Q}$). That is, there are ten 'parallel classes' of lines, each class containing nine (mutually parallel) lines.

If we adjoin ten new points, each corresponding to a parallel class (and lying on every line in that class), and if we also adjoin a new line (comprising the ten new points), then the enlarged system is a projective plane of order 9: this plane we call Ω.

Denote the new point corresponding to a parallel class of nine lines $y = x\mu + \kappa$ by (μ); and the new point corresponding to the parallel class of lines $x = \lambda$ by Y. If we write $X = (0)$ then X and Y may be likened to the 'points at infinity' on the x- and y-axes in the Euclidean plane.

Summarizing: the points of Ω are

81 'proper' points (x, y) $(x, y \in \mathcal{Q})$

9 'ideal' points (μ) $(\mu \in \mathcal{Q})$

1 'ideal' point Y

and the lines are

81 'proper' lines $y = x\mu + \kappa$ $(\mu, \kappa \in \mathcal{Q})$

9 'proper' lines $x = \lambda$ $(\lambda \in \mathcal{Q})$

1 'ideal' line XY (where $X = (0)$) having no equation.

The ideal point (μ) lies on $y = x\mu + \kappa$, for all $\kappa \in \mathcal{Q}$, and on XY. The ideal point Y lies on $x = \lambda$, for all $\lambda \in \mathcal{Q}$, and on XY.

The equation of the line joining points (x_1, y_1) and (x_2, y_2), with $x_1 \neq x_2$ and $y_1 \neq y_2$, is

$$(y - y_2)(y_1 - y_2)^{-1} = (x - x_2)(x_1 - x_2)^{-1}.$$

There is, because $\mathscr{Q}$ is not left-distributive, no corresponding form for the coordinates of the point of intersection of two lines $y = x\mu_1 + \kappa_1$ and $y = x\mu_2 + \kappa_2$. Herein lies the main difficulty of the computations needed to explore the geometry of Ω.

4.2 Some collineations of Ω

The operations of addition and multiplication in $\mathscr{Q}$ determine respectively two types of collineations: addition gives the 81 transformations

$$\mathscr{T}_{\delta,\epsilon}\colon (x, y) \to (x+\delta, y+\epsilon) \quad (\delta, \epsilon \in \mathscr{Q}),$$

while multiplication gives the 64 transformations

$$\mathscr{S}_{\rho,\sigma}\colon (x, y) \to (x\rho, y\sigma) \quad (\rho, \sigma \in \mathscr{Q};\ \rho, \sigma \neq 0).$$

EXERCISE 4.2.1 Show that $\mathscr{T}_{\delta,\epsilon}$ and $\mathscr{S}_{\rho,\sigma}$ are 1–1 transformations of the set of proper points in Ω onto itself.

Note that since multiplication in $\mathscr{Q}$ is not commutative, it is important to distinguish between ρx and $x\rho$. Indeed

$$(x, y) \to (\rho x, \sigma y)$$

does not in general give a collineation. For instance, if $\rho = \sigma = \alpha$, then the three collinear points $(0, 1)$, $(1, -1)$, (α, γ) (all on the line $y = x+1$) are mapped to the three non-collinear points $(0, \alpha)$, $(\alpha, -\alpha)$, $(-1, -\beta)$ (the first two, but not the third, of which lie on $y = x+\alpha$).

The transformations $\mathscr{T}_{\delta,\epsilon}$ and $\mathscr{S}_{\rho,\sigma}$ were defined on the proper points (x, y) only. Under any one of these mappings proper points go to proper points, and every proper point is the image of a proper point. It follows that if lines are mapped to lines then pairs of parallel lines are mapped to pairs of parallel lines, and the transformations extend in a natural fashion to the ideal points. So to show that the transformations extend to collineations we need only to show that proper lines are mapped to proper lines.

Consider $\mathscr{T}_{\delta,\epsilon}$ first. Suppose the point (x_0, y_0) lies on the line $y = x\mu + \kappa$. Then

$$y_0 = x_0\mu + \kappa.$$

Therefore
$$y_0+\epsilon = x_0\mu+\kappa+\epsilon$$
$$= (x_0+\delta)\,\mu+\kappa+\epsilon-\delta\mu.$$

So $(x_0+\delta, y_0+\epsilon)$ lies on the line $y = x\mu+(\kappa+\epsilon-\delta\mu)$. Similarly if (x_0, y_0) lies on $x = \lambda$ then $(x_0+\delta, y_0+\epsilon)$ lies on $x = \lambda+\delta$. $\mathscr{T}_{\delta,\epsilon}$ fixes each of the ideal points (since a line through (μ) or Y is mapped to a line through (μ) or Y, respectively).

Now consider $\mathscr{S}_{\rho,\sigma}$. If $y_0 = x_0\mu+\kappa$ then
$$y_0\sigma = (x_0\mu+\kappa)\,\sigma$$
$$= x_0\mu\sigma+\kappa\sigma$$
$$= (x_0\rho)\,(\rho^{-1}\mu\sigma)+\kappa\sigma.$$

So $(x_0\rho, y_0\sigma)$ lies on $y = x(\rho^{-1}\mu\sigma)+\kappa\sigma$. If (x_0, y_0) lies on $x = \lambda$ then $(x_0\rho, y_0\sigma)$ lies on $x = \lambda\rho$. $\mathscr{S}_{\rho,\sigma}$ maps the ideal point (μ) to $(\rho^{-1}\mu\sigma)$, while Y is fixed.

The transformation $\mathscr{R} : (x, y) \to (y, x)$ (in the Euclidean plane, this formula defines the reflection in the line $y = x$) also extends to a collineation of Ω. Once again, proper points are mapped, one-to-one, to proper points. If (x_0, y_0) lies on $y = x\mu+\kappa$, with $\mu \neq 0$, then
$$y_0\mu^{-1} = (x_0\mu+\kappa)\,\mu^{-1}$$
$$= x_0+\kappa\mu^{-1}.$$

Therefore $x_0 = y_0\mu^{-1}-\kappa\mu^{-1}$; that is, (y_0, x_0) lies on
$$y = x\mu^{-1}-\kappa\mu^{-1}.$$

The line $y = \kappa$ is mapped to the line $x = \kappa$, and vice versa. $\mathscr{R}$ maps (μ), with $\mu \neq 0$, to (μ^{-1}), while interchanging $X = (0)$ and Y.

$\mathscr{U}$: $(x, y) \to (x+y, x-y)$ extends to a collineation. Thus if $y_0 = x_0\mu+\kappa$, where $\mu \in \mathscr{Q}^*$ (so that $\mu^2 = -1$, $\mu^{-1} = -\mu$), then
$$x_0-y_0 = x_0-x_0\mu-\kappa$$
$$= (x_0-x_0\mu)\,\mu\mu^{-1}-\kappa$$
$$= (x_0\mu+x_0)\,\mu^{-1}-\kappa$$
$$= (y_0-\kappa+x_0)\,\mu^{-1}-\kappa$$
$$= (x_0+y_0)\,\mu^{-1}-\kappa-\kappa\mu^{-1}$$
$$= (x_0+y_0)\,(-\mu)-\kappa+\kappa\mu,$$

so that (x_0+y_0, x_0-y_0) lies on $y = x(-\mu)-\kappa+\kappa\mu$. If $y_0 = x_0+\kappa$ then $x_0-y_0 = -\kappa$; that is, (x_0+y_0, x_0-y_0) lies on $y = -\kappa$; and if $y_0 = -x_0+\kappa$ then $x_0+y_0 = \kappa$, so that (x_0+y_0, x_0-y_0) lies on $x = \kappa$. The line $y = \kappa$ goes to the line $y = x+\kappa$ and the line $x = \kappa$ goes to the line $y = -x-\kappa$. $\mathscr{U}$ maps (0) to (1), Y to (-1), (1) to (0), (-1) to Y, and (μ) to $(-\mu)$ when $\mu \neq 0, \pm 1$.

Finally, let $\mathscr{S}$ be any automorphism of $\mathscr{Q}$. Then

$$\mathscr{S}(\rho+\sigma) = \mathscr{S}(\rho)+\mathscr{S}(\sigma) \quad \text{and} \quad \mathscr{S}(\rho\sigma) = \mathscr{S}(\rho)\mathscr{S}(\sigma)$$

for all $\rho, \sigma \in \mathscr{Q}$, so that

$$y = x\mu+\kappa \Leftrightarrow \mathscr{S}(y) = \mathscr{S}(x)\,\mathscr{S}(\mu)+\mathscr{S}(\kappa) \quad (\mathscr{S}^{-1} \in \operatorname{Aut}(\mathscr{Q}) \text{ also})$$

and the line with equation $y = x\mu+\kappa$ is mapped to the line $y = x\mathscr{S}(\mu)+\mathscr{S}(\kappa)$ when $(x, y) \to (\mathscr{S}(x), \mathscr{S}(y))$. Also the line $x = \tau$ is mapped to $x = \mathscr{S}(\tau)$. When we extend to the ideal points, $(\mu) \to (\mathscr{S}(\mu))$ and $Y \to Y$.

NOTATION 4.2.1 $\mathscr{A}_\lambda$: the collineation of Ω determined by the automorphism of $\mathscr{Q}$ which maps α to λ.

There are six collineations $\mathscr{A}_\lambda$, one corresponding to each element λ of $\mathscr{Q}^*$ (cf. Theorem 1.4.1).

Summarizing, we have:

THEOREM 4.2.1 *The following are collineations of* Ω:

(i) $\mathscr{T}_{\delta,\epsilon}$: $(x, y) \to (x+\delta, y+\epsilon)$, $(\mu) \to (\mu)$, $Y \to Y$.

(ii) $\mathscr{S}_{\rho,\sigma}$: $(x, y) \to (x\rho, y\sigma)$, $(\mu) \to (\rho^{-1}\mu\sigma)$, $Y \to Y$ $(\rho, \sigma \neq 0)$.

(iii) $\mathscr{R}$: $(x, y) \to (y, x)$, $(\mu) \to (\mu^{-1})$ *if* $\mu \neq 0$, $X \to Y$, $Y \to X$.

(iv) $\mathscr{U}$: $(x, y) \to (x+y, x-y)$, $(0) \to (1)$, $(1) \to (0)$,

$(-1) \to Y$, $Y \to (-1)$, $(\mu) \to (-\mu)$ if $\mu \neq 0, \pm 1$.

(v) $\mathscr{A}_\lambda$: $(x, y) \to (\mathscr{S}x, \mathscr{S}y)$, $(\mu) \to (\mathscr{S}\mu)$, $Y \to Y$, *where*

$$\mathscr{S} \in \operatorname{Aut}(\mathscr{Q}) \quad \textit{and} \quad \mathscr{S}(\alpha) = \lambda \in \mathscr{Q}^*.$$

EXERCISE 4.2.2 Show that, under composition, the set of all collineations $\mathscr{T}_{\delta,\epsilon}$ forms a commutative group and the set of all collineations $\mathscr{S}_{\rho,\sigma}$ a non-commutative group.

EXERCISE 4.2.3 Show that $\mathscr{U}^2 = \mathscr{S}_{-1,-1}$, $\mathscr{U}^4 = 1$.

If an ideal line is adjoined to the real Euclidean plane, the pairs of ideal points belonging to the pairs of perpendicular parallel line-classes form a special family: each congruence transformation, since it preserves perpendicularity, permutes the pairs in this family.

On the line XY of the plane Ω there is a similar family of pairs of points: one whose pairs are permuted by all collineations. (In the Euclidean plane, there are collineations which are not congruence transformations and do not preserve the special pairs.) The five special pairs are:

$$X = (0), Y; \quad (1), (-1); \quad (\alpha), (-\alpha); \quad (\beta), (-\beta); \quad (\gamma), (-\gamma).$$

DEFINITION 4.2.1 The five pairs $\{X, Y\}$, $\{(\mu), (-\mu)\}$, $\mu = 1, \alpha, \beta, \gamma$, of ideal points in Ω are called *special pairs*.

EXERCISE 4.2.4 Show that each of the collineations of Theorem 4.2.1 permutes the five special pairs.

We conclude this section by proving some results which are to be used repeatedly in this chapter.

THEOREM 4.2.2 *If $\mu \neq 0$, there is a collineation which maps (μ) to X and $(-\mu)$ to Y.*

$\mathscr{U}$ maps (1) to (0) $= X$. Also $\mathscr{U}$ maps (-1) to Y while $\mathscr{R}$ maps Y to X, so $\mathscr{R}\mathscr{U}$ maps (-1) to X. Finally if $\mu \neq 0, \pm 1$, $\mathscr{S}_{\mu,1}$ maps (μ) to $(\mu^{-1}.\mu.1) = (1)$, and $\mathscr{U}$ maps (1) to X, so $\mathscr{U}\mathscr{S}_{\mu,1}$ maps (μ) to X. By Exercise 4.2.4 any product of collineations $\mathscr{S}_{\rho,\sigma}$, $\mathscr{R}$, $\mathscr{U}$ which maps (λ), $\lambda \neq 0$, to X necessarily maps $(-\lambda)$ to Y.

THEOREM 4.2.3 *If $\mu \neq 0$ there is a collineation which fixes X and Y and maps (μ) to (1).*

$\mathscr{S}_{\mu,1}$ is the required collineation.

EXERCISE 4.2.5 $\{V_1, W_1\}$, $\{V_2, W_2\}$ are distinct special pairs, as are also $\{V_3, W_3\}$, $\{V_4, W_4\}$. Show that there is a collineation which maps V_1 to V_3, W_1 to W_3, V_2 to V_4, W_2 to W_4.

THEOREM 4.2.4 *Given two distinct proper points A, B and two distinct proper points A', B' (in Ω) there is a collineation which maps A to A', B to B'.*

By Theorem 4.2.2 (or if $AB \cap XY = Y$, by using the collineation $\mathscr{R}$) there is a collineation which maps $AB \cap XY$ to X. Suppose this collineation maps A to $A_1 = (\rho, \sigma)$, B to B_1. Then the collineation $\mathscr{T}_{-\rho,-\sigma}$ maps A_1 to $(0,0)$ and B_1 to say B_2, while fixing X. $(0,0)$, B_2, X are collinear, being the images of the collinear points A, B, $AB \cap XY$ under a collineation. So $B_2 = (\tau, 0)$ for some $\tau \neq 0$. The collineation $\mathscr{S}_{\tau^{-1},1}$ fixes $(0,0)$ and maps B_2 to $(1,0)$. Combining all these collineations we obtain a collineation which maps A to $(0,0)$, B to $(1,0)$. Similarly we may find a collineation which maps $(0,0)$ to A', $(1,0)$ to B'.

NOTATION 4.2.2 $O = (0,0), \quad I = (1,1), \quad X = (0).$

EXERCISE 4.2.6 Show, using Theorem 4.2.4 and the collineations $\mathscr{A}_\sigma$, that any three distinct collinear proper points A, B, C can be mapped (in order) to either O, I, $(-1,-1)$ or O, I, (α, α).

4.3 The collineation group of Ω

Each collineation $\mathscr{T}_{\delta,\epsilon}$ fixes every point of XY, and therefore, being an axial collineation, is also a central collineation (Exercise 2.2.2). $\mathscr{T}_{\delta,\epsilon}$ maps (x_0, y_0) to $(x_0+\delta, y_0+\epsilon)$ and the join of these points is the line $y - y_0 = (x - x_0)\,\delta^{-1}\epsilon$ if $\delta \neq 0$, or $x = x_0$ if $\delta = 0$. Thus, for any proper point H, $\langle H, \mathscr{T}_{\delta,\epsilon} H\rangle$ is parallel to $y = x\delta^{-1}\epsilon$ or, when $\delta = 0$, to $x = 0$, and:

THEOREM 4.3.1 *$\mathscr{T}_{\delta,\epsilon}$ is an elation with the ideal line as axis; when $\delta \neq 0$ the centre is $(\delta^{-1}\epsilon)$ and when $\delta = 0$ the centre is Y.*

Applying Theorem 2.2.4, we deduce:

THEOREM 4.3.2 *Ω is (W, XY)-transitive for all W on XY; that is, Ω is a translation plane with respect to XY.*

The elations $\mathscr{T}_{\delta,\epsilon}$, with axis XY, are often called 'the translations of Ω'.

The collineations $\mathscr{S}_{\rho,\sigma}$ are not in general central, but we are to prove:

Theorem 4.3.3 *$\mathscr{S}_{\rho,1}$ is an (X, OY)-homology, and $\mathscr{S}_{\rho,\sigma}$ is a projectivity.*

Take $H = (x_0, y_0)$, and $H' = \mathscr{S}_{\rho,1}H = (x_0\rho, y_0)$. $\mathscr{S}_{1,1} = 1$, so suppose $\rho \neq 1$. Then $H' = H$ if and only if $x_0 = 0$. Also $\mathscr{S}_{\rho,1}$ fixes Y. So $\mathscr{S}_{\rho,1}$ has as axis the line $x = 0$. The centre is X since HH' (when $H' \neq H$) has equation $y = y_0$.

Exercise 4.3.1 Prove that $\mathscr{S}_{1,\sigma}$ is a (Y, OX)-homology.

The second part of Theorem 4.3.3 now follows from the relation $\mathscr{S}_{\rho,1}\mathscr{S}_{1,\sigma} = \mathscr{S}_{\rho,\sigma}$.

The existence of the homologies $\mathscr{S}_{\rho,1}$ and $\mathscr{S}_{1,\sigma}$ implies, by Theorem 2.2.4:

Theorem 4.3.4 *Ω is (X, OY)-transitive and (Y, OX)-transitive.*

Given any proper point P, there is a $\mathscr{T}_{\delta,\epsilon}$ which maps O to P, while fixing all the ideal points. It follows, by Theorem 2.3.3, that:

Theorem 4.3.5 *Ω is (X, PY)-transitive and (Y, PX)-transitive, for every proper point P.*

The pair $\{X, Y\}$ may be mapped to any other given special pair of ideal points (Exercise 4.2.5). Consequently:

Theorem 4.3.6 *Ω is (W, PW')-transitive, for all special pairs W, W' of ideal points, and all proper points P.*

Exercise 4.3.2 Show that $\mathscr{R}$ is an (E', OE)-homology, where $E = (1), E' = (-1)$.

We are to prove next:

Theorem 4.3.7 *If l is any proper line and L is the ideal point on l, there is no non-trivial (L, l)-elation.*

We consider first the possibility of an (X, OX)-collineation, treating separately the cases in which the ideal point (1) is mapped to Y, (-1), and (μ) where $\mu \in \mathcal{Q}^*$. (1) cannot here be mapped to X, and only the trivial (X, OX)-collineation maps (1) to itself. Assume there is an (X, OX)-collineation $\mathscr{C}$, and that

$$\mathscr{C}(x_0, y_0) = (x_0', y_0).$$

The construction for (x_0', y_0) is shown in Diagram 4.3.

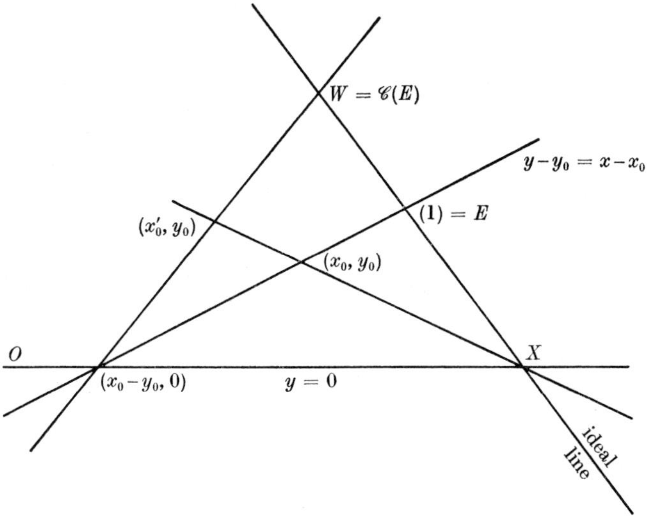

Diagram 4.3

(*a*) $(1) \to Y$: $x_0' = x_0 - y_0$

$$\mathscr{C}\{(0,0), (1, \alpha), (\alpha, -1)\} = \{(0,0), (-\beta, \alpha), (\gamma, -1)\}.$$

The original points are collinear whereas the points to which they are mapped are not.

(*b*) $(1) \to (-1)$: $x_0' = x_0 + y_0$

$$\mathscr{C}\{(0,0), (1, \alpha), (\alpha, -1)\} = \{(0,0), (\gamma, \alpha), (\beta, -1)\}.$$

The three image points are non-collinear.

(*c*) $(1) \to (\mu)$, $\mu \in \mathcal{Q}^*$.

Because the collineations $\mathscr{A}_\lambda$ fix both X and OX we may (cf. Theorem 2.2.3) suppose that $\mu = \alpha$. Then $x_0' = x_0 - y_0 - y_0\alpha$ and

$$\mathscr{C}\{(0,0,), (1, -1), (-\alpha, \alpha)\} = \{(0,0), (\beta, -1), (\gamma, \alpha)\}.$$

Again three collinear points are mapped to three non-collinear points. Thus there is no non-trivial (X, OX)-collineation in Ω.

If there existed a non-trivial (L, l)-collineation $\mathscr{B}$ then, since there is a collineation $\mathscr{V}$ which maps L to X and l to OX, there would be a non-trivial (X, OX)-collineation $\mathscr{V}\mathscr{B}\mathscr{V}^{-1}$.

Since Ω is not (X, OX)-transitive, the plane is non-Desarguesian. In particular:

THEOREM 4.3.8 *Ω is not isomorphic to the Galois plane Φ.*

The next step towards the identifying of all central collineations is provided by the following Theorem:

THEOREM 4.3.9 *There is no non-trivial (V, PW)-collineation in Ω, if V and W are distinct ideal points not forming a special pair, and P is a proper point.*

The proof can be completed in the following steps: there is no non-trivial (L, m)-collineation if (i) $L = X$, $m = OE$, (ii) $L = X$, $m = OW$, (iii) $L = V$, $m = OW$, (iv) $L = V$, $m = PW$.

We turn now to collineations of which the centre is not an ideal point. One of these can be identified immediately, namely $\mathscr{S}_{-1,-1}$: $(x, y) \to (-x, -y)$, which clearly has O as centre.

EXERCISE 4.3.3 Prove that $\mathscr{S}_{-1,-1}$ is an (O, XY)-collineation and, except for the identity, the only such collineation.

EXERCISE 4.3.4 Let P be any proper point. Prove that there is one and only one non-trivial (P, XY)-collineation.

The following theorem completes our determination of the central collineations in Ω.

THEOREM 4.3.10 *There is no non-trivial (P, l)-collineation for which P is a proper point and l is a proper line.*

Suppose $\mathscr{C}$ is such a collineation. Let m be the (proper) line to which $\mathscr{C}$ maps the ideal line, and let L be the ideal point on l. Then $L = l \cap m = \mathscr{C}(L)$. Also, since L is ideal, Ω is (L, XY)-transitive. But $\mathscr{C}(L) = L$ and $\mathscr{C}(XY) = m$. Therefore, by Theorem 2.2.3, Ω is (L, m)-transitive. This is impossible, by Theorem 4.3.7, since L is the ideal point on m.

We bring together and summarize some of the preceding results in:

THEOREM 4.3.11 *A collineation of Ω is central if and only if it may be expressed as a conjugate $\mathscr{X}\mathscr{C}\mathscr{X}^{-1}$ of a collineation $\mathscr{C}$, with*

$$\mathscr{C} = \mathscr{T}_{\delta,\epsilon}, \quad \mathscr{S}_{\rho,1} \quad or \quad \mathscr{S}_{-1,-1}$$

and $\mathscr{X} \in \mathfrak{G}$, where $\mathfrak{G} = \langle \mathscr{R}, \{\mathscr{S}_{\rho,\sigma}\}, \{\mathscr{T}_{\delta,\epsilon}\}, \mathscr{U}, \{\mathscr{A}_\lambda\} \rangle$ is the group generated by the collineations listed in Theorem 4.2.1.

There is a fairly simple criterion for deciding whether a given collineation of Ω is or is not projective. We consider the action of the various collineations on the ten ideal points. First we prove:

THEOREM 4.3.12 *Every collineation of Ω maps the ideal line onto itself.*

For suppose there is a collineation which maps XY to some proper line l, mapping X to say P. Then, since Ω is (X, XY)-transitive, it is also (P, l)-transitive. But $P \in l$ and l is proper, so we have a contradiction to Theorems 4.3.7 and 4.3.10.

Now consider the permutations of the set of ten ideal points induced by the collineations $\mathscr{C}$ of Theorem 4.3.11. The collineations $\mathscr{T}_{\delta,\epsilon}$ and $\mathscr{S}_{-1,-1}$ all fix each ideal point, that is induce the identity permutation, while $\mathscr{S}_{\rho^{-1},1}$ induces the permutation

$$(X, Y, (1), (-1), (\alpha), (-\alpha), (\beta), (-\beta), (\gamma), (-\gamma))$$
$$\to (X, Y, (\rho), (-\rho), (\rho\alpha), (-\rho\alpha), (\rho\beta), (-\rho\beta), (\rho\gamma), (-\rho\gamma)).$$

For each value of ρ it is easily verified that this is an even permutation. Thus, in every case, $\mathscr{C}$ induces an even permutation. It follows that every central collineation induces an even permutation of the ideal points, since $\mathscr{X}\mathscr{C}\mathscr{X}^{-1}$ gives an even permutation if $\mathscr{C}$ does, whether or not $\mathscr{X}$ induces an even permutation. But projectivities are products of central collineations. Thus:

THEOREM 4.3.13 *Every projective collineation of Ω induces an even permutation on the set of ten ideal points.*

This was first proved by J. André (1955), who also proved that the non-projective collineations induce odd permutations on the ideal points.

EXERCISE 4.3.5 (i) Show that $\mathscr{U}$ is not a projectivity.

(ii) Prove that, from a given point A, the point $C = \mathscr{U}A$ can be constructed in the following way: Construct

$$H = YA \cap OX, \quad B = HE \cap XA, \quad K = XB \cap OY,$$

where $E = (1)$; then $C = KE \cap YB$.

We conclude this section by showing that the group $\mathfrak{G}$ is the whole collineation group of Ω; that is:

THEOREM 4.3.14 $\mathfrak{G} = \langle \mathscr{R}, \{\mathscr{S}_{\rho,\sigma}\}, \{\mathscr{T}_{\delta,\epsilon}\}, \mathscr{U}, \{A_\lambda\}\rangle$ *is the entire collineation group of* Ω.

Consider an arbitrary collineation $\mathscr{X}$ of Ω. $\mathscr{X}$ maps the ideal line onto itself, and (by Theorems 4.3.6 and 4.3.9) permutes the five special pairs of ideal points. Now there exist:

(i) $\mathscr{A} \in \mathfrak{G}$ such that $\mathscr{A}\mathscr{X}$ fixes the ideal points X and Y (Theorem 4.2.2),

(ii) $\mathscr{B} \in \{\mathscr{T}_{\delta,\epsilon}\}$ such that $\mathscr{B}\mathscr{A}\mathscr{X}$ fixes X, Y and O,

(iii) $\mathscr{C} \in \{\mathscr{S}_{\rho,\sigma}\}$ such that $\mathscr{C}\mathscr{B}\mathscr{A}\mathscr{X}$ fixes X, Y, O and I,

(iv) $\mathscr{D} \in \{\mathscr{A}_\lambda\}$ such that $\mathscr{D}\mathscr{C}\mathscr{B}\mathscr{A}\mathscr{X}$ fixes X, Y, O, I and (α).

The collineation $\mathscr{Y} = \mathscr{D}\mathscr{C}\mathscr{B}\mathscr{A}\mathscr{X}$ fixes the points X, Y, O, I, whose completion is easily seen to be a subplane of order 3 containing the points (x, y) with $x, y \in \mathscr{D}$ and the four ideal points $X = (0), (1), (-1), Y$. Therefore $\mathscr{Y}$ fixes every point of this subplane. But the points fixed by $\mathscr{Y}$ form a subplane: the line joining two fixed points is a fixed line, the intersection of two fixed lines is a fixed point, and there is a regular quadrangle O, I, X, Y of fixed points. Also Y fixes the point (α). So the fixed points of $\mathscr{Y}$ form a subplane of order greater than 3. By Bruck's Theorem, this subplane must be the whole plane Ω; that is, $\mathscr{Y} = 1$ and

$$\mathscr{X} = \mathscr{A}^{-1}\mathscr{B}^{-1}\mathscr{C}^{-1}\mathscr{D}^{-1} \in \mathfrak{G}.$$

EXERCISE 4.3.6 Coordinates in Ω can be expressed in the quasi-homogeneous forms:

$$(x, y, 1) \quad \text{for proper points,}$$

$$(x, 1, 0) \quad \text{and} \quad (1, 0, 0) \quad \text{for ideal points.}$$

Also, since $(x, y, 0)$ and $(x, 0, 0)$ can be used without ambiguity instead of $(y^{-1}x, 1, 0)$ and $(1, 0, 0)$, we may write all three forms together as (x, y, e) where $e = 0$ or 1. Prove that the collineations which belong to the subgroup $\mathfrak{H} = \langle \mathscr{R}, \{\mathscr{S}_{\rho,\sigma}\}, \{\mathscr{T}_{\delta,\epsilon}\}, \mathscr{U} \rangle$ of $\mathfrak{G}$ can be expressed in the form

$$[x, y, e] \rightarrow [x, y, e]\,\mathbf{M},$$

where $\mathbf{M}$ is a product of matrices

$$\mathbf{T}_{\delta,\epsilon} = \begin{bmatrix} 1 & 0 & 0 \\ 0 & 1 & 0 \\ \delta & \epsilon & 1 \end{bmatrix}, \quad \mathbf{S}_{\rho,\sigma} = \begin{bmatrix} \rho & 0 & 0 \\ 0 & \sigma & 0 \\ 0 & 0 & 1 \end{bmatrix},$$

$$\mathbf{R} = \begin{bmatrix} 0 & 1 & 0 \\ 1 & 0 & 0 \\ 0 & 0 & 1 \end{bmatrix}, \quad \mathbf{U} = \begin{bmatrix} 1 & 1 & 0 \\ 1 & -1 & 0 \\ 0 & 0 & 1 \end{bmatrix}.$$

EXERCISE 4.3.7 Prove that $\mathbf{R}$ can be expressed as a product of matrices $\mathbf{U}$ and $\mathbf{S}_{-1,-1}$ (so that, in Theorem 4.3.14, $\mathscr{R}$ is redundant).

EXERCISE 4.3.8 (i) Prove that, given $\rho, \sigma, \delta, \epsilon$, we can find δ', ϵ' such that

$$\mathscr{T}_{\delta,\epsilon}\mathscr{S}_{\rho,\sigma} = \mathscr{S}_{\rho,\sigma}\mathscr{T}_{\delta',\epsilon'}.$$

(ii) Prove that $\mathscr{U}$ commutes with all $\mathscr{T}_{\delta,\epsilon}$, and that $\mathscr{A}_\lambda$ commutes with all $\mathscr{S}_{\rho,\sigma}$, with all $\mathscr{T}_{\delta,\epsilon}$ and with $\mathscr{U}$.

(iii) Prove that every collineation of Ω can be expressed in the form

$$\mathscr{A}_\lambda \mathscr{T}_{\delta,\epsilon}\mathscr{S}_{\rho,\sigma}\mathscr{U}\mathscr{S}_{\rho',\sigma'}\mathscr{U}\mathscr{S}_{\rho'',\sigma''}\mathscr{U}\ \ldots.$$

4.4 Fano subplanes of Ω

A regular quadrangle $ABCD$ completes to a Fano plane if and only if the diagonal points, $BC \cap AD$, $CA \cap BD$, $AB \cap CD$, are collinear. Consider the four points $O = (0, 0)$, $I = (1, 1)$, X, Y in Ω. $OI \cap XY = (1)$, $OX \cap IY = (1, 0)$, $OY \cap IX = (0, 1)$. These three points are not collinear. Now take instead of O, I, X, Y the four points O, I, X, P, where P is any point on OY except O, Y, and $OY \cap IX$; then $OIXP$ is a regular quadrangle.

$$P = (0, \rho) \quad \text{with} \quad \rho \neq 0, 1.$$

We find the diagonal points:

$$OI \cap XP = (\rho, \rho), \quad OX \cap IP = (-\rho(1-\rho)^{-1}, 0)$$

since IP has equation $(y-\rho)(1-\rho)^{-1} = (x-0)(1-0)^{-1}$, and

$$OP \cap IX = (0, 1).$$

The line joining the first and last of these three diagonal points has equation

$$(y-1)(\rho-1)^{-1} = (x-0)(\rho-0)^{-1},$$

so that the three points are collinear if and only if

$$-(\rho-1)^{-1} = -\rho(1-\rho)^{-1}\rho^{-1};$$

that is,
$$\rho - 1 = \rho(1-\rho)\rho^{-1}.$$

If $\rho = -1$ then $\rho - 1 = 1$ while $\rho(1-\rho)\rho^{-1} = -1$. But if

$$\rho \in \mathscr{Q}^* = \{\pm\alpha, \pm\beta, \pm\gamma\}$$

then, since $\beta = \alpha - 1$ and $\gamma = \alpha + 1$, $1 - \rho \in \mathscr{Q}^*$ also and therefore, as $\rho \neq \pm(1-\rho)$,

$$\rho(1-\rho) = -(1-\rho)\rho$$

so that

$$\rho(1-\rho)\rho^{-1} = -(1-\rho)\rho\rho^{-1} = -(1-\rho) = \rho - 1.$$

We have proved that if $\rho \neq 0, \pm 1$ then O, I, X, $(0, \rho)$ lie in a Fano subplane. The remaining three points of the Fano subplane are the three diagonal points.

The proof of the following theorem is contained in Exercises 4.4.1 and 4.4.2.

THEOREM 4.4.1 *Every Fano subplane of Ω contains precisely one ideal point. There are just six Fano subplanes containing O, I, X, namely the completions of the regular quadrangles O, I, X, $(0, \rho)$, with $\rho \in \mathscr{Q}^*$.*

Exercise 4.4.1 will establish the second statement.

Let P be a fourth point determining a Fano subplane containing O, I, X: that is, $OIXP$ is a regular quadrangle with collinear diagonal points. There is only one such point P in each Fano subplane containing O, I, X, since each of the lines OI, IX, XO contains a third point of the subplane.

EXERCISE 4.4.1 (*a*) Show that P is not ideal.

Hint. Let $P = (\mu)$ and consider separately the cases $\mu = 0$, $\mu = 1, \mu = -1, \mu \in \mathcal{Q}^*$. ($P = Y$ was excluded above.)

Now suppose that none of the seven lines of the subplane passes through Y. P is proper: let $P = (\rho, \sigma)$.

EXERCISE 4.4.1 (*b*) Show that $\rho \neq 0, 1$; $\sigma \neq 0, 1$; $\rho \neq \sigma$.

EXERCISE 4.4.1 (*c*) Show that the three diagonal points are:

$$OI \cap XP = (\sigma, \sigma), \quad OX \cap IP = (1-(\sigma-1)^{-1}(\rho-1), 0),$$
$$OP \cap IX = (\sigma^{-1}\rho, 1).$$

These three diagonal points lie on a line d, and $d \not\ni Y$.

EXERCISE 4.4.1 (*d*) Show that $\rho \in \mathcal{Q}^*$ and $\sigma \in \mathcal{Q}^*$.

Hint. Show that (i) if $\rho = -1$ then $\sigma \in \mathcal{Q}^*$ and d passes through Y, (ii) if $\sigma = -1$ then $\rho \in \mathcal{Q}^*$ and the diagonal points are not collinear.

Since $\rho \in \mathcal{Q}^*$ and $\sigma \in \mathcal{Q}^*$,

$$1-(\sigma-1)^{-1}(\rho-1) = 1+(\sigma-1)(\rho-1),$$

and

$$\sigma^{-1}\rho = -\sigma\rho.$$

Thus $OI \cap XP = (\sigma, \sigma), \quad OX \cap IP = (1+(\sigma-1)(\rho-1), 0),$

$$OP \cap IX = (-\sigma\rho, 1).$$

EXERCISE 4.4.1 (*e*) Show that $\rho\sigma = -\sigma\rho$.

Hint. Verify that if $\rho = -\sigma$ then the diagonal points are not collinear (assuming $\rho \in \mathcal{Q}^*$), which gives a contradiction.

Since the line d passes through (σ, σ) and $(-\sigma\rho, 1)$ it has equation $(y-\sigma)(1-\sigma)^{-1} = (x-\sigma)(-\sigma\rho-\sigma)^{-1}$; that is

$$(y-\sigma)(\sigma-1) = (x-\sigma)(\rho\sigma-\sigma)^{-1}.$$

Now $(\rho\sigma-\sigma)^{-1} = [(\rho-1)\sigma]^{-1} = \sigma^{-1}(\rho-1)^{-1} = \sigma(\rho-1)$. So d has equation $(y-\sigma)(\sigma-1) = (x-\sigma)\sigma(\rho-1)$, that is

$$(y-\sigma)(\sigma-1) = (x\sigma+1)(\rho-1).$$

Substituting $(1+(\sigma-1)(\rho-1), 0)$ for (x, y), we derive

$$-\sigma(\sigma-1) = \{[1+(\sigma-1)(\rho-1)]\sigma+1\}(\rho-1),$$

then $$(\sigma-1)\sigma = [\sigma+(\sigma-1)(\rho-1)\sigma+1](\rho-1),$$

or $$\begin{aligned} -(\sigma+1) &= [\sigma-(\rho-1)(\sigma-1)\sigma+1](\rho-1) \\ &= [\sigma+1+(\rho-1)(\sigma+1)](\rho-1) \\ &= \rho(\sigma+1)(\rho-1). \end{aligned}$$

EXERCISE 4.4.1 (*f*) Show the impossibility of the last equation when

(i) $\sigma = \rho+1$, (ii) $\sigma = -(\rho+1)$, (iii) $\sigma = \rho-1$, (iv) $\sigma = -(\rho-1)$.

(These are the only relations between ρ and σ not previously excluded.)

It follows that each of the Fano subplanes containing O, I, X has a line through Y. By our earlier work there are only six such.

We have so far only attempted to construct a Fano subplane to contain a regular quadrangle with one ideal vertex. To show that every Fano subplane contains an ideal point, the reader should work the following sequence of exercises.

Let Γ be a Fano subplane whose seven points are proper.

EXERCISE 4.4.2 (*a*) The ideal points on the seven lines of Γ include at least two of the five special pairs, say $\{L, L'\}$ and $\{M, M'\}$.

EXERCISE 4.4.2 (*b*) If A is the intersection of the lines of Γ through L and L', and B is the intersection of the lines through M and M', then $A \neq B$.

EXERCISE 4.4.2 (*c*) If no three of AL, AL', BM, BM' are concurrent then:

(i) there is a collineation which maps L to X, L' to Y, M to (1), M' to (-1), A to $(0, 0)$ and B to $(1, \sigma)$, for some $\sigma \in \mathcal{Q}^*$;

(ii) the points of Γ on AL' map to $(0, 0)$, $(0, \sigma-1)$, $(0, \sigma+1)$, those on AL to $(0, 0)$, $(1-\sigma, 0)$, $(1+\sigma, 0)$;

(iii) the point of Γ not yet determined is ideal, giving a contradiction.

EXERCISE 4.4.2 (*d*) If $A \in BM'$ then:

(i) there is a collineation which maps L to X, L' to Y, M to (1), M' to (-1), A to $(0, 0)$ and B to $(-1, 1)$;

(ii) $AL \cap BM$ maps to $(1, 0)$, $AL' \cap BM$ to $(0, -1)$;

(iii) the third point of Γ on AL maps to $(\rho, 0)$ for some $\rho \in \mathcal{Q}^*$;

(iv) the three diagonal points of the quadrangle $(0, 0)$, $(-1, 1)$, $(0, -1)$, $(\rho, 0)$ are not collinear, so that Γ is not a Fano subplane, contradicting the definition of Γ.

EXERCISE 4.4.2 (*e*) The other possibilities left by Exercise 4.4.2 (*c*), namely $A \in BM$, $B \in AL$, $B \in AL'$, are all similar to the case $A \in BM'$.

Thus every Fano subplane has at least one ideal point. In an arbitrary Fano subplane, let V be an ideal point, paired with V', and choose two proper points P and Q such that V, $V' \notin PQ$. Then P, Q, V can be mapped to O, I, X, the Fano subplane mapping to another Fano subplane. The latter contains only one ideal point by Exercise 4.4.1, and so the former contains only one ideal point. This completes the proof of the theorem.

THEOREM 4.4.2 *If A and B are two proper points, W an ideal point, paired with W', and AB contains neither W nor W', then there are precisely six Fano subplanes which contain A, B and W and each has a line which passes through W'.*

Map A, B, W to O, I, X and apply Theorem 4.4.1.

EXERCISE 4.4.3 Verify that the number of Fano subplanes in Ω is $81.64.10 = 51840$.

4.5 Subplanes of order 3 in Ω

There are ten ideal points in Ω. The points (x, y) with

$$x, y \in \mathcal{D} = \{0, 1, -1\}$$

belong to a subplane Δ_0 of order 3, obtained by adjoining the four ideal points $X = (0)$, (1), (-1), Y.

THEOREM 4.5.1 *Every subplane of order 3 in Ω includes four of the ten ideal points.*

A subplane Δ of order 3 includes 13 of the 91 lines in Ω. If one of these 13 is the ideal line then the four points of Δ on this line are ideal. Suppose on the other hand that all 13 lines are proper. Then each meets the ideal line in a single point. There are only ten ideal points, so two of these 13 lines meet the ideal line in the same point P_1; P_1 must lie in Δ, and on four of the 13 lines of Δ. Either each of the remaining nine ideal points lies on one of the remaining nine lines; or one of these ideal points, say P_2, lies on two of the lines, that is the ideal line ($= P_1P_2$) is in Δ contrary to our assumption.

If all thirteen lines are proper, four of them pass through one ideal point, say W, and the other nine pass through each of the other ideal points. Let W' be the ideal point paired with W, let K, L, M, N be the four points of Δ on the line of Δ through W', and let H be one of the two other points of Δ on WK. We can

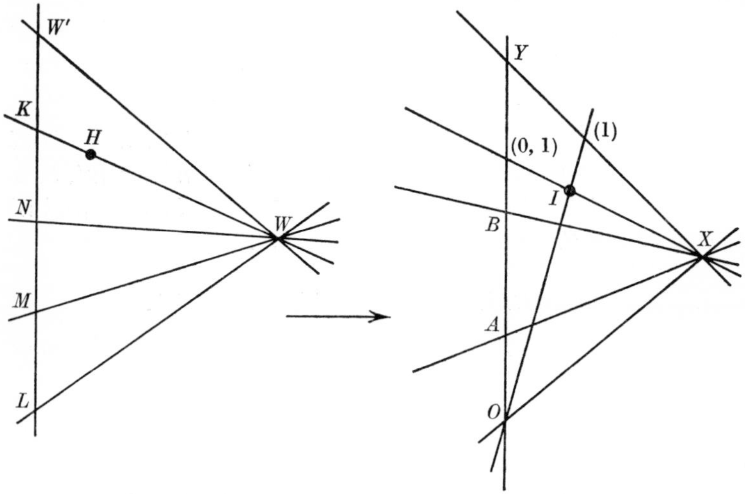

Diagram 4.5

find a collineation in which $W \to X$, $W' \to Y$, $N \to O$, $H \to I$ and say $L \to A$, $M \to B$.

Since O, I, X, $(0, \rho)$ generate a Fano subplane whenever $\rho \in \mathcal{Q}^*$, A and B must be two of the points $O, (0, \pm 1)$. Thus our hypothesis that the thirteen lines are all proper has led to the impossible conclusion that either $A = (0, 1)$ or $B = (0, 1)$ or $A = O$ or $B = O$ or $A = B$. So every subplane of order 3 contains four ideal points.

EXERCISE 4.5.1 Show that there is a subplane Δ_1, of order 3, containing O, I, X, (α), and that in it the four ideal points are X, (1), (α), $(-\beta)$.

We now prove:

THEOREM 4.5.2 *Δ_0 is the real subplane, and Δ_1 is the subplane which is the completion of the quadrangle O, I, X, (α). If Δ is any other subplane of order* 3 *then either* (i) *the four ideal points of Δ comprise two of the special pairs, in which case there is a collineation which maps Δ to Δ_0; or* (ii) *the four ideal points of Δ include none of the special pairs, in which case there is a collineation which maps Δ to Δ_1.*

(i) Suppose Δ contains a special pair. There is a collineation which maps this pair to X, Y, and Δ to say Δ'. Let A, B be two proper points in Δ', such that AB passes through neither X nor Y. There is a collineation $\mathscr{T}_{\delta,\epsilon}$ which maps A to O, and a collineation $\mathscr{S}_{\rho,\sigma}$ which fixes O and maps $B' = \mathscr{T}_{\delta,\epsilon} B$ to I. $\mathscr{S}_{\rho,\sigma}\mathscr{T}_{\delta,\epsilon}$ maps Δ' to a subplane Δ'' of order 3 containing O, I, X, Y. Since Δ_0 is also a subplane of order 3 containing O, I, X, Y, $\Delta'' = \Delta_0$.

(ii) Suppose Δ contains no special pair. Let the ideal points of Δ be U_1, U_2, U_3, U_4. There is a collineation $\mathscr{H}$ which maps U_1 to X, U_2 to $E = (1)$. Let A, B be proper points of Δ collinear with U_2. Then $A' = \mathscr{H}A$, $B' = \mathscr{H}B$ and (1) are collinear. Let $\mathscr{T}_{\delta,\epsilon}$ map A' to O, B' to say B''. $\mathscr{T}_{\delta,\epsilon}$ fixes all ideal points. Now $B'' \in OE$, so there is a collineation $\mathscr{S}_{\rho,\rho}$ fixing X, E and O, and mapping B'' to I. $\mathscr{S}_{\rho,\rho}\mathscr{T}_{\delta,\epsilon}\mathscr{H}$ maps Δ to a subplane Δ', containing O, I, X, E. Let U_3', U_4' be the images of U_3, U_4. Since no two of X, (1), U_3', U_4' are paired we may write $U_3' = (\sigma)$, where $\sigma \in \mathscr{Q}^*$. So the collineation $\mathscr{A}_\sigma^{-1}$ corresponding to the automorphism of $\mathscr{Q}$ which maps σ to α, fixes O, I, X and maps $U_3' = (\sigma)$ to (α); that is, $\mathscr{A}_\sigma^{-1}$ maps Δ' to Δ_1.

EXERCISE 4.5.2 Show that in Ω there are $5.8.9$ subplanes of order 3 equivalent to Δ_0 and $5.16.9$ equivalent to Δ_1, a total of 1080.

EXERCISE 4.5.3 What are the possible intersections of (i) two subplanes of order 3 equivalent to Δ_0, (ii) two subplanes equivalent to Δ_1, (iii) a subplane equivalent to Δ_0 with a subplane equivalent to Δ_1?

4.6 An oval and some constructions in Ω

In §2.6 conics were defined in the plane over an arbitrary commutative field. A non-singular conic in a Galois plane of order q consists of $q+1$ points, no three collinear. B. Segre (1955) has shown that, when q is odd, every such set of points is a conic. This leads us to consider similarly defined sets of points in other planes.

DEFINITION 4.6.1 An *oval* in a finite plane of order n is a set of $n+1$ points, the *vertices* of the oval, such that no three are collinear.

Since the joins of any vertex P to the other vertices are a set of n lines, there is one other line through P, and this line contains only one vertex, P. As in the case of a non-singular conic, this line is called the *tangent* at P.

G. Rodriguez (1959) discovered that a set of points which maps to the following set of ten points forms an oval in Ω:

$$\mathcal{O}_1 = \{X, Y, (1,1), (-1,-1), (\beta,-\beta), (-\beta,\beta), (\gamma,\alpha), (-\gamma,-\alpha), (\alpha,\gamma), (-\alpha,-\gamma)\}.$$

We shall see that $\mathcal{O}_1$, while it determines tangents (and therefore exterior and interior points) and a symmetric relation between exterior points and chords, does not determine a polarity of Ω. In fact, as we shall prove in §4.7, Ω is not self-dual, and can therefore in particular have no polarities.

EXERCISE 4.6.1 (*a*) Verify (i) that the vertices of $\mathcal{O}_1$ are permuted by the collineations $\mathcal{R}, \mathcal{S}_{\beta,-\beta}, \mathcal{S}_{\alpha,\gamma}$ and $\mathcal{A}_{-\gamma}$, and (ii) that the eight proper vertices of $\mathcal{O}_1$ are cyclically permuted by the collineation $\mathcal{L} = \mathcal{S}_{\alpha,\gamma}\mathcal{A}_{-\gamma}$.

EXERCISE 4.6.1 (*b*) (i) By calculating the slopes of the lines joining (1, 1) to the remaining seven proper vertices of $\mathcal{O}_1$, show that no line through (1, 1) contains more than two vertices of $\mathcal{O}_1$.

(ii) Verify that no line through X contains more than two vertices of $\mathcal{O}_1$.

(iii) Deduce that $\mathcal{O}_1$ is an oval.

EXERCISE 4.6.2 Verify that every chord of $\mathcal{O}_1$ can be transformed, by a collineation in the group $\langle \mathcal{R}, \mathcal{S}_{\beta,-\beta}, \mathcal{S}_{\alpha,\gamma}, \mathcal{A}_{-\gamma}\rangle$, to one of the five chords $XY, XP_0, P_0P_4, P_0P_2, P_0P_1$, where

$$P_0 = (1, 1) \quad \text{and} \quad P_r = \mathcal{L}^r P_0;$$

and that the numbers of different chords to which these five chords can be transformed are respectively 1, 16, 4, 8, 16. What is the order of the group?

EXERCISE 4.6.3 By considering the effect of the collineations $\mathcal{S}_{\alpha,\alpha}, \mathcal{S}_{\beta,\beta}, \mathcal{S}_{\gamma,\gamma}$ show that the 32 proper points other than

$$O = (0, 0)$$

on the lines OP_0, OP_1, OP_2, OP_3 can be split into four sets of eight, each set, together with X, Y, forming an oval.

The reason for our choice of the particular transform $\mathcal{O}_1$ of the Rodriguez oval is that if, as in § 1.3, we write $\alpha = \omega$, $\beta = \omega^6$, $\gamma = \omega^7$, then $P_r = (\omega^r, \omega^{-r})$ for all r; that is:

THEOREM 4.6.1 *The set of ten points*

$$\mathcal{O}_1 = \{(\omega^r, \omega^{-r}) : r = 0, \ldots, 7\} \cup \{X, Y\}$$

is an oval in Ω.

This formula, with suitable ranges for r, may be used also to obtain ovals in planes over certain near-fields of orders other than 9.

In Table 4.6.1 we set out:

(i) The names and coordinates of the vertices of $\mathcal{O}_1$ in the first two rows and last two columns.

(ii) The names and equations of the tangents of $\mathcal{O}_1$ in the first two columns and last two rows. The tangent at P_r is t_r. The equation of, for example, t_7, to which in the second column corresponds the entry '$\beta, -\alpha$', is $y = x\beta - \alpha$.

(iii) In the lower left-hand triangle, the coordinates of the intersections of pairs of tangents; for example, $t_5 \cap t_0 = (\gamma, -\beta)$.

(iv) In the upper right-hand triangle the equations of the chord —for example, P_5P_0 is $y = x\alpha - \beta$.

There are four chords through each exterior point, and four exterior points on each chord.

EXERCISE 4.6.4 Verify that if $H = t_i \cap t_j$ and $h = P_iP_j$ are an exterior point and its associated chord, and $\{H_r\}$, $r = 1, \ldots, 4$, is the set of exterior points on a chord, and $\{h_r\}$ is the set of associated chords, then:

(i) If the chord is XY, P_0P_4 or P_0P_2, the four lines h_r are concurrent.

(ii) If the chord is XP_0 or P_0P_1 then only three of the lines h_r concur, but not all four.

Thus, since the correspondence between exterior points and their associated chords is not such that four collinear points always correspond to four concurrent lines, this correspondence cannot be induced by a correlation.

We conclude this section with an investigation of the adaptations to Ω of two constructions for conics, and various constructions associated with triangles, in the Euclidean plane. We shall find, with the first two constructions, that a set of ten points is obtained, but that in one case six points are collinear and in the other eight.

The first construction is the analogue of the DTP construction of Definition 2.6.1.

EXERCISE 4.6.5 $XYAB$ is a fixed regular quadrangle in Ω, and $XY \cap AB = E$, $XA \cap YB = F$, $XB \cap YA = G$. l is a variable line through B and $M = XY \cap l$, $N = YA \cap l$, $P = NX \cap AM$.

Table 4.6.1 *Vertices, tangents, chords and exterior points of the Rodriguez oval*

		Vertices											
		X	Y	P_0	P_1	P_2	P_3	P_4	P_5	P_6	P_7		
Tangents		.	.	$1, 1$	α, γ	$-\beta, \beta$	$-\gamma, -\alpha$	$-1, -1$	$-\alpha, -\gamma$	$\beta, -\beta$	γ, α		
OX	$y = 0$	.	XY	$0, 1$	$0, \gamma$	$0, \beta$	$0, -\alpha$	$0, -1$	$0, -\gamma$	$0, -\beta$	$0, \alpha$	.	X
OY	$x = 0$	$0, 0$	.	$x = 1$	$x = \alpha$	$x = -\beta$	$x = -\gamma$	$x = -1$	$x = -\alpha$	$x = \beta$	$x = \gamma$	.	Y
t_0	$-1, -1$	$-1, 0$	$0, -1$	.	$\gamma, -\alpha$	$-\beta, \alpha$	$-\alpha, \gamma$	$1, 0$	$\alpha, -\beta$	$\beta, -\gamma$	$-\gamma, \beta$	$1, 1$	P_0
t_1	$-\beta, -\gamma$	$-\alpha, 0$	$0, -\gamma$	$\beta, -\alpha$	.	$-\gamma, -1$	$1, 1$	α, β	$\beta, 0$	$-\alpha, \alpha$	$-1, -\beta$	α, γ	P_1
t_2	$1, -\beta$	$\beta, 0$	$0, -\beta$	$-\gamma, \alpha$	$\gamma, -1$	.	$\gamma, -\gamma$	β, γ	$-\alpha, -\alpha$	$-1, 0$	$\alpha, 1$	$-\beta, \beta$	P_2
t_3	β, α	$\gamma, 0$	$0, \alpha$	$-\beta, \gamma$	$-1, 1$	$1, -\gamma$	.	$-\gamma, -\beta$	$-1, \beta$	$\alpha, -1$	$\beta, 0$	$-\gamma, -\alpha$	P_3
t_4	$-1, 1$	$1, 0$	$0, 1$	(-1)	$-\gamma, \beta$	$-\alpha, \gamma$	$\alpha, -\beta$	.	γ, α	$-\beta, -\alpha$	$-\alpha, -\gamma$	$-1, -1$	P_4
t_5	$-\beta, \gamma$	$\alpha, 0$	$0, \gamma$	$\gamma, -\beta$	$(-\beta)$	$-1, -\alpha$	β, β	$-\beta, \alpha$	.	$-\alpha, -\alpha$	$1, -1$	$-\alpha, -\gamma$	P_5
t_6	$1, \beta$	$-\beta, 0$	$0, \beta$	$\alpha, -\gamma$	$1, \alpha$	(1)	$-\alpha, -1$	$\gamma, -\alpha$	$-\gamma, 1$	.	γ, γ	$\beta, -\beta$	P_6
t_7	$\beta, -\alpha$	$-\gamma, 0$	$0, -\alpha$	$-\alpha, \beta$	$-\beta, -\beta$	$\alpha, 1$	(β)	$\beta, -\gamma$	$1, -1$	$-1, \gamma$	.	γ, α	P_7
		$y = 0$ OX	$x = 0$ OY	$-1, -1$ t_0	$-\beta, -\gamma$ t_1	$1, -\beta$ t_2	β, α t_3	$-1, 1$ t_4	$-\beta, \gamma$ t_5	$1, \beta$ t_6	$\beta, -\alpha$ t_7		

Prove that the ten points P are the four vertices of the regular quadrangle X, Y, A, $EG \cap YB$ together with the six points on EF other than $E, F, EF \cap XB$ and $EF \cap YA$. (Take $A = (0,0)$, $B = (1,1)$.)

The second construction is the analogue of the construction of a circle as the locus of the point of intersection of two perpendicular lines drawn one through each of two fixed points. ('The angle in a semi-circle is a right angle.') Some of the non-metrical properties of perpendicularity in the Euclidean plane may be summed up in the statements:

(i) $a \perp b \Rightarrow b \perp a$,

(ii) $\{a\colon a \perp a\} = \varnothing$,

(iii) $a \perp b$ and $b \perp c \Rightarrow a \| c$,

(iv) a single line can be drawn through any point perpendicular to any line.

If we define perpendicularity in Ω thus: '$a \perp b$ if and only if the ideal points on a, b form a special pair', then we obtain a relation which has all four of the properties listed. This leads us to:

Exercise 4.6.6 A, B are two fixed proper points of Ω, W is the ideal point on AB and $\{W, W'\}$ is a special pair. $\{V, V'\}$ is any other special pair and $P = AV \cap BV'$. Prove that the eight points P constructible in this way lie on a line, say k, and that the other two points on k are W' and $k \cap AB$. (Take $A = (0,0)$, $B = (1,0)$, $\{W, W'\} = \{X, Y\}$.)

The 'circle' consists of A, B and the eight points P on k!

Since arithmetic in $\mathscr{Q}$ is modulo 3, we might define a 'mid-point' in the following way: the *mid-point* of the point-pair $\{(x_1, y_1), (x_2, y_2)\}$ is $(-(x_1+x_2), -(y_1+y_2))$.

Exercise 4.6.7 Show that each of the collineations $\mathscr{R}$, $\mathscr{S}_{\rho,\sigma}$, $\mathscr{T}_{\delta,\epsilon}$, $\mathscr{U}$, $\mathscr{A}_\lambda$ maps the mid-point of two (proper) points to the mid-point of their images.

The line $x = -1$ which contains eight points of the 'circle' in Exercise 4.6.6 is perpendicular to AB ($y = 0$). In fact, since $(0,0)+(1,0) = (1,0)$, $(-1,0)$ is the mid-point of $\{A, B\}$ and the line $x = -1$ might be called the 'right bisector' of $\{A, B\}$. So the 'circle' consists of A, B and the eight proper points on the 'right bisector' of $\{A, B\}$ which do not lie on AB.

We consider now the 'medians' of a triangle ABC (A, B, C non-collinear proper points in Ω). The three ideal points of AB, BC, CA either (i) belong to three different special pairs or (ii) two of them belong to the same pair.

EXERCISE 4.6.8 Show that in case (i) there is a collineation which maps A to (0, 0), B to (1, 0) and C to $(\rho, 1-\rho)$, $\rho \in \mathcal{Q}^*$; and that in case (ii) there is a collineation which maps A to $(0, 0)$, B to $(1, 0)$ and C to $(0, 1)$.

The mid-point of $\{(0, 0), (1, 0)\}$ is $(-1, 0)$ and the line joining this to $(\rho, 1-\rho)$ is $y = -(x+1)\rho$; the mid-point of $\{(1, 0), (\rho, 1-\rho)\}$ is $(-\rho-1, \rho-1)$ and the line joining this to $(0, 0)$ is $y = -x\rho$; the mid-point of $\{(\rho, 1-\rho), (0, 0)\}$ is $(-\rho, \rho-1)$ and the line joining this to $(1, 0)$ is $y = -(x-1)\rho$. So the 'medians' of a triangle of type (i) are concurrent, but the point of concurrency is ideal!

EXERCISE 4.6.9 Show that the 'medians' of a triangle of type (ii) concur at an ideal point.

EXERCISE 4.6.10 Are the 'altitudes' of a triangle in Ω necessarily concurrent?

EXERCISE 4.6.11 Are the 'right bisectors' of the 'sides' of a triangle in Ω necessarily concurrent?

4.7 The dual plane Ω^D

The plane Ω may be described as follows:

Points		Lines	
(ρ, σ)	$(\rho, \sigma \in \mathcal{Q})$	$[\mu, \kappa]$	$(\mu, \kappa \in \mathcal{Q})$
(τ)	$(\tau \in \mathcal{Q})$	$[\lambda]$	$(\lambda \in \mathcal{Q})$
Y		XY	

Incidence conditions

(i) $(\rho, \sigma) \in [\mu, \kappa] \Leftrightarrow \sigma = \rho\mu + \kappa$,

$(\rho, \sigma) \in [\lambda] \Leftrightarrow \rho = \lambda$,

$(\rho, \sigma) \notin XY$.

(ii) $(\tau) \in [\mu, \kappa] \Leftrightarrow \tau = \mu$,

$(\tau) \notin [\lambda]$ but $(\tau) \in XY$.

(iii) $Y \notin [\mu, \kappa]$ but $Y \in [\lambda]$ and $Y \in XY$.

The dual plane Ω^D is obtained by taking as points the lines of Ω, as lines the points of Ω, and retaining the same incidence conditions (cf. §2.1). Thus the relation

$\sigma = \rho\mu + \kappa$ has two interpretations:

in Ω, the point (ρ, σ) lies on the line $[\mu, \kappa]$;

in Ω^D, the point (μ, κ) lies on the line $[\rho, \sigma]$.

That is, the equation of the line $[\mu, \kappa]$ in Ω is $y = x\mu + \kappa$, and the equation of the line $[\rho, \sigma]$ in Ω^D is $\sigma = \rho x + y$, or $y = -\rho x + \sigma$.

If we give the line $[\rho, \sigma]$ in Ω^D a new label $[-\rho, \sigma]$ then we can say 'the line $[-\rho, \sigma]$ in Ω^D has equation $y = -\rho x + \sigma$' or equivalently 'the line $[\rho, \sigma]$ in Ω^D has equation $y = \rho x + \sigma$'. To retain the incidence condition $\rho = \lambda$ for a line $[\rho, \sigma]$ and a point (λ) in Ω^D, we relabel the point (λ) as $(-\lambda)$. Then the list of incidence conditions for Ω^D is the same as that for Ω except that $\sigma = \rho\mu + \kappa$ is replaced by $\sigma = \mu\rho + \kappa$; that is, whereas in Ω the lines have equations with coefficients on the right, for Ω^D coefficients are on the left.

Each result obtained for Ω gives rise to a dual result for Ω^D. For example, Theorem 4.3.1, that Ω is (W, XY)-transitive for all points W on the line XY, gives us the theorem: Ω^D is (Y, w)-transitive for all lines w through the point Y. Since there is no point P in Ω such that Ω is (P, l)-transitive for all lines l through P (Theorems 4.3.7 and 4.3.10) the plane Ω^D is distinct from Ω, that is:

THEOREM 4.7.1 *Ω is not self-dual.*

We give a list of some corresponding results for Ω and Ω^D.

	Ω	Ω^D
1	Ω is (W, XY)-transitive for all W on XY	Ω^D is (Y, w)-transitive for all w through Y
2	In Ω the ten points on XY can be split into five pairs so that every collineation maps any pair to another pair (permutes the pairs)	In Ω^D the ten lines through Y can be split into five pairs so that every collineation maps any pair to another pair (permutes the pairs)
3	If $\{L, L'\}$ is a special pair, Ω is (L', l)-transitive for all lines l through L	If $\{l, l'\}$ is a special pair, Ω^D is (L, l')-transitive for all points L on l
4	Each Fano subplane of Ω contains exactly one point of XY	Each Fano subplane of Ω^D contains exactly one line through Y
5	Four of the thirteen points of any subplane of order 3 in Ω lie on XY	Four of the thirteen lines of any subplane of order 3 in Ω^D pass through Y

EXERCISE 4.7.1 Using the theorem that the set of ten points

$$\{(x, y, 1): x = y^2 \quad \text{and} \quad x, y \in \mathscr{F}\} \cup \{(1, 0, 0)\}$$

is an oval in Φ, show that

$$\mathscr{O}_2 = \{(\omega^{2r}, \omega^r): r = 0, \ldots, 7\} \cup \{(0, 0), X\}$$

is an oval in Ω^D.

Chapter 5

THE PLANE Ψ

5.1 The construction and the collineation group of Ψ

In the system Ψ, points are to be represented by homogeneous coordinate-vectors $\boldsymbol{\lambda}$ with miniquaternion components. A clear distinction will be drawn between 'real points' and 'complex points': we shall prove that there is no collineation which maps a real point to a complex point. In this property Ψ is quite different from Ω, since in Ω the collineation $\mathscr{T}_{\delta,\epsilon}$, for example, maps real points to complex points if δ or ϵ is complex, although the ideal line is always 'real'.

Definition 5.1.1 (i) *The point* $\boldsymbol{\lambda}$ is the set of vectors $\{\boldsymbol{\lambda}\kappa : \kappa \in \mathscr{Q}, \kappa \neq 0\}$. (It is always assumed that $\boldsymbol{\lambda} \neq \mathbf{0}$.)

(ii) *Real point*: a point $\boldsymbol{\lambda}$ such that, for some $\kappa \neq 0$, $\boldsymbol{\lambda}\kappa$ is a real vector (that is, all three of its components belong to $\mathscr{D}$).

(iii) *Complex point*: any point that is not a real point.

The numbers of real and complex points depend only on the numbers of distinct sets $\{\boldsymbol{\lambda}\kappa\}$ and are therefore the same as the corresponding numbers for Φ:

Theorem 5.1.1 *In* Ψ *there are* 13 *real points and* 78 *complex points.*

We propose the following definition of a line in Ψ and shall prove that it satisfies all the conditions necessary to establish that Ψ is a projective plane.

Definition 5.1.2 A *line*, $\langle \mathbf{p}, \boldsymbol{\lambda} \rangle$, of Ψ is the set of ten points $\{\mathbf{p}\} \cup \{\mathbf{p}\kappa + \boldsymbol{\lambda} : \kappa \in \mathscr{Q}\}$, where $\mathbf{p}$ is a fixed real point, and $\boldsymbol{\lambda}$ is a fixed point distinct from $\mathbf{p}$.

It is necessary to verify that $\langle \mathbf{p}, \boldsymbol{\lambda} \rangle$ depends only on the points $\mathbf{p}$ and $\boldsymbol{\lambda}$, and not on the choice of coordinate-vectors for these points; that is, that we obtain the same set of points when $\mathbf{p}$ and $\boldsymbol{\lambda}$ are replaced by $\mathbf{p}\theta$ and $\boldsymbol{\lambda}\phi$ ($\theta, \phi \neq 0$). We have

$$(\mathbf{p}\theta)\kappa + \boldsymbol{\lambda}\phi = (\mathbf{p}(\theta\kappa\phi^{-1}) + \boldsymbol{\lambda})\phi$$

and thus the sets of points

$$\{\mathbf{p}\kappa + \boldsymbol{\lambda} : \kappa \in \mathscr{Q}\} \quad \text{and} \quad \{(\mathbf{p}\theta)\kappa + \boldsymbol{\lambda}\phi : \kappa \in \mathscr{Q}\}$$

are identical.

We are using the notation $\langle \mathbf{p}, \boldsymbol{\lambda} \rangle$ in anticipation of the result, which we must be careful not to assume at this stage, that there is a *unique* line joining $\mathbf{p}$ and $\boldsymbol{\lambda}$.

From Definition 5.1.2 every line has at least one real point, and we separate real from complex lines by the following definition:

DEFINITION 5.1.3 *Real line*: a line which contains at least two real points. *Complex line*: a line which contains only one real point.

Before considering the system of real points and real lines we note that, if $\boldsymbol{\mu}$ is a point (distinct from $\mathbf{p}$) of the line $\langle \mathbf{p}, \boldsymbol{\lambda} \rangle$, then $\langle \mathbf{p}, \boldsymbol{\lambda} \rangle = \langle \mathbf{p}, \boldsymbol{\mu} \rangle$. For let $\boldsymbol{\mu} = \mathbf{p}\phi + \boldsymbol{\lambda}$; then

$$\begin{aligned} \langle \mathbf{p}, \boldsymbol{\mu} \rangle &= \{\mathbf{p}\kappa + (\mathbf{p}\phi + \boldsymbol{\lambda}) : \kappa \in \mathscr{Q}\} \cup \{\mathbf{p}\} \\ &= \{\mathbf{p}(\kappa + \phi) + \boldsymbol{\lambda} : \kappa \in \mathscr{Q}\} \cup \{\mathbf{p}\} \quad \text{since } \mathbf{p} \text{ is real} \\ &= \langle \mathbf{p}, \boldsymbol{\lambda} \rangle. \end{aligned}$$

THEOREM 5.1.2 *The real points and the real lines of Ψ form a projective plane Δ_0 of order* 3.

The real points on a real line $\langle \mathbf{p}, \mathbf{q} \rangle = \{\mathbf{p}\kappa + \mathbf{q} : \kappa \in \mathscr{Q}\} \cup \{\mathbf{p}\}$, where $\mathbf{q}$ is real, are the four points $\mathbf{p}, \mathbf{q}, \mathbf{p} \pm \mathbf{q}$. There is an equation $a_1x_1 + a_2x_2 + a_3x_3 = 0$ ($a_i \in \mathscr{D}$, not all zero) satisfied by $\mathbf{x} = \mathbf{p}$ and $\mathbf{x} = \mathbf{q}$. The solutions of this equation, over $\mathscr{D}$, are

$$\mathbf{x} = \mathbf{p}a + \mathbf{q}b,$$

for all real a, b. So the real points and real lines form a plane over the field $\mathscr{D}$; that is, a plane of order 3.

Note that, if the real vectors $\mathbf{p}$ and $\mathbf{q}$ satisfy $\Sigma a_i x_i = 0$, with $a_i \in \mathscr{D}$, then $\mathbf{p}\kappa + \mathbf{q}$ also satisfies the equation for all κ, real or complex. Thus the real line $\langle \mathbf{p}, \mathbf{q} \rangle$ has the same equation whether considered as a line of Ψ or as a line of Δ_0.

We are to prove that there is a unique line joining any two distinct points of Ψ. The line joining two real points is necessarily real, and so, by Theorem 5.1.2, unique. If $\mathbf{p}$ is a real point and $\boldsymbol{\lambda}$ a complex point, then $\langle \mathbf{p}, \boldsymbol{\lambda} \rangle$ is a line joining $\mathbf{p}$ and $\boldsymbol{\lambda}$. Suppose $\langle \mathbf{q}, \boldsymbol{\mu} \rangle$ is another. If $\mathbf{q} = \mathbf{p}$ (as points) then $\langle \mathbf{q}, \boldsymbol{\mu} \rangle$ contains $\boldsymbol{\lambda}$; that is

$$\langle \mathbf{q}, \boldsymbol{\mu} \rangle = \langle \mathbf{q}, \boldsymbol{\lambda} \rangle = \langle \mathbf{p}, \boldsymbol{\lambda} \rangle;$$

whereas if $\mathbf{q} \neq \mathbf{p}$ then

$$\langle \mathbf{q}, \boldsymbol{\mu} \rangle = \langle \mathbf{q}, \mathbf{p} \rangle = \langle \mathbf{p}, \mathbf{q} \rangle$$

contains $\boldsymbol{\lambda}$ and so

$$\langle \mathbf{q}, \boldsymbol{\mu} \rangle = \langle \mathbf{p}, \mathbf{q} \rangle = \langle \mathbf{p}, \boldsymbol{\lambda} \rangle.$$

The case of two complex points is more difficult We shall use certain mappings determined by the projectivities $\mathbf{x} \to \mathbf{Mx}$ ($\mathbf{M}$ a non-singular 3×3 real matrix) of Δ_0. The matrix $\mathbf{M}$ may be used to define a mapping $\mathscr{M}$ of the system Ψ onto itself, if we write also $\mathbf{x} \to \mathbf{Mx}$ for complex $\mathbf{x}$. This mapping permutes the 91 points of Ψ, and also the lines of Ψ: for

$$\mathbf{p}\kappa + \boldsymbol{\lambda} \to \mathbf{M}(\mathbf{p}\kappa + \boldsymbol{\lambda}) = \mathbf{Mp}\kappa + \mathbf{M}\boldsymbol{\lambda}$$

(since $\mathbf{M}$ is real), so that $\langle \mathbf{p}, \boldsymbol{\lambda} \rangle \to \langle \mathbf{Mp}, \mathbf{M}\boldsymbol{\lambda} \rangle$.

THEOREM 5.1.3 *There is a unique line joining any two distinct points* $\boldsymbol{\lambda}$ *and* $\boldsymbol{\mu}$ *of* Ψ.

We may suppose that $\boldsymbol{\lambda}$ and $\boldsymbol{\mu}$ are complex points, since the other cases were dealt with above. There is a unique real line through the complex point $\boldsymbol{\lambda}$: since the intersection of two real lines is a real point, there is at most one; and $\boldsymbol{\lambda} + \boldsymbol{\lambda}^*$, $\boldsymbol{\lambda} - \boldsymbol{\lambda}^*$ are real points such that $\boldsymbol{\lambda} \in \langle \boldsymbol{\lambda} + \boldsymbol{\lambda}^*, \boldsymbol{\lambda} - \boldsymbol{\lambda}^* \rangle$. (Here, as in Exercise 1.4.3, $(a + \beta b)^* = a - \beta b$ if a and b are real.)

Suppose $\boldsymbol{\lambda}$ and $\boldsymbol{\mu}$ lie on the same real line $\langle \mathbf{p}, \mathbf{q} \rangle$. Using the mappings $\mathscr{M}$, we may take $\mathbf{p} = (1, 0, 0)$ and $\mathbf{q} = (0, 1, 0)$. Then $\langle \mathbf{p}, \mathbf{q} \rangle$ has equation $x_3 = 0$, and so $\lambda_3 = 0$, $\mu_3 = 0$. If $\boldsymbol{\mu} \in \langle \mathbf{r}, \boldsymbol{\lambda} \rangle$, then $\mathbf{r}\kappa + \boldsymbol{\lambda} = \boldsymbol{\mu}\theta$ for some κ, θ. Thus $r_3 = 0$ and so $\langle \mathbf{r}, \boldsymbol{\lambda} \rangle$ has equation $x_3 = 0$; that is, $\langle \mathbf{r}, \boldsymbol{\lambda} \rangle = \langle \mathbf{p}, \mathbf{q} \rangle$.

Suppose, on the other hand, that $\boldsymbol{\lambda}$ and $\boldsymbol{\mu}$ lie, respectively, on two distinct real lines l and m. Using the mappings $\mathscr{M}$ again, we may take these lines to be $x_2 = 0$ and $x_3 = 0$, and write $\boldsymbol{\lambda} = (\lambda, 0, 1)$, $\boldsymbol{\mu} = (\mu, 1, 0)$, with λ, μ complex. The line $\langle \mathbf{r}, \boldsymbol{\nu} \rangle$ contains $\boldsymbol{\lambda}$ and $\boldsymbol{\mu}$ if and only if $\langle \mathbf{r}, \boldsymbol{\nu} \rangle = \langle \mathbf{r}, \boldsymbol{\lambda} \rangle$ and $\mathbf{r}\kappa + \boldsymbol{\lambda} = \boldsymbol{\mu}\theta$ for some non-zero κ, θ. The last equation is equivalent to the three equations: $r_1\kappa + \lambda = \mu\theta$, $r_2\kappa = \theta$, $r_3\kappa + 1 = 0$. Since r_1, r_2, r_3 are to be real, κ, θ and $\lambda - \mu\theta$ must be real. Now λ and μ are complex, so $\mu = \pm\lambda + b$ (b real), and $\lambda - \mu\theta$ is real if and only if either $\mu = \lambda + b$ and $\theta = 1$ or $\mu = -\lambda + b$ and $\theta = -1$; that is, θ is uniquely determined by λ and μ. It follows that $\mathbf{r}\kappa$ ($= \boldsymbol{\mu}\theta - \boldsymbol{\lambda}$) is uniquely determined by $\boldsymbol{\lambda}$ and $\boldsymbol{\mu}$, and there is exactly one line $\langle \mathbf{r}, \boldsymbol{\lambda} \rangle$ containing $\boldsymbol{\lambda}$ and $\boldsymbol{\mu}$.

We have now completed the proof that there is a unique line joining two points of Ψ. It is easily verified that (1, 0, 0), (0, 1, 0), (0, 0, 1), (1, 1, 1) is a regular quadrangle. The simple counting argument set out in Exercise 5.1.1 below shows that any two lines of Ψ have a common point, so that:

THEOREM 5.1.4 *Ψ is a projective plane of order* 9.

EXERCISE 5.1.1 Show that any two lines of Ψ have a common point, by using the properties: (i) there are ten points on each line, (ii) there is exactly one line joining two distinct points, and (iii) there are exactly 91 lines in Ψ. (Prove the last assertion.)

We return now to the mappings $\mathscr{M}: \mathbf{x} \to \mathbf{Mx}$, associated with the non-singular real matrices $\mathbf{M}$. Since $\mathscr{M}$ is a permutation of the points which carries lines onto lines, $\mathscr{M}$ is a collineation of Ψ. This collineation is an extension of the projectivity $\mathbf{x} \to \mathbf{Mx}$, $\mathbf{x}$ real, of the real subplane Δ_0. In fact $\mathscr{M}$ is a projectivity of the plane Ψ, since $\mathbf{M}$ is a product of the matrices representing homologies and elations of Δ_0, and these matrices give also homologies and elations in Ψ. In the case of a homology in Δ_0, the matrix is similar over $\mathscr{D}$ to a matrix

$$\mathbf{H} = \begin{bmatrix} h & 0 & 0 \\ 0 & 1 & 0 \\ 0 & 0 & 1 \end{bmatrix};$$

and the corresponding collineation $\mathscr{H}$ of Ψ fixes all ten points on $x_1 = 0$ and all ten lines through $(1, 0, 0)$. The elations also extend to elations by an analogous argument.

THEOREM 5.1.5 *The mappings* $\mathbf{M}: \mathbf{x} \to \mathbf{Mx}$, $\mathbf{M}$ *real and non-singular, are projectivities of* Ψ.

NOTATION 5.1.1 ∇: the group of projectivities $\mathscr{M}$.

THEOREM 5.1.6 *The group* ∇ *is transitive on*

(i) *the incident real point-line pairs* (R, r),
(ii) *the incident (real point)–(complex line) pairs* (R, c),
(iii) *the incident (complex point)–(real line) pairs* (C, r),
(iv) *the incident complex point–line pairs* (C, c).

To prove this theorem we use the following lemma:

LEMMA *A necessary and sufficient condition that the line* $\langle \mathbf{p}, \boldsymbol{\lambda} \rangle$, *where* $\boldsymbol{\lambda} = \boldsymbol{l} + \beta \boldsymbol{l}'$ *and* $\boldsymbol{l}, \boldsymbol{l}'$, *are real vectors, should be real is that* $\det [\mathbf{p}, \boldsymbol{l}, \boldsymbol{l}'] = 0$.

First we express multiplication in the form

$$\rho\sigma = (r + \beta r')(s + \beta s') = rs - j_\sigma r's' + \beta(rs' + j_\sigma r's),$$

where
$$j_\sigma = 1 \quad \text{when} \quad \sigma \in \{\pm 1, \pm \beta\},$$
$$j_\sigma = -1 \quad \text{when} \quad \sigma \in \{\pm \alpha, \pm \gamma\}.$$

(Cf. Exercise 1.3.5.)

Now if $\kappa = k + \beta k' \neq 0$ and $\mathbf{p}(h + \beta h') + (\boldsymbol{l} + \beta \boldsymbol{l}')(k + \beta k')$ is a real vector, then

$$\mathbf{p}h' + \boldsymbol{l}k' + j_\kappa \boldsymbol{l}'k = \mathbf{0}.$$

Conversely, if
$$\mathbf{p}a + \boldsymbol{l}b + j_\sigma \boldsymbol{l}'c = \mathbf{0},$$

then

$$\mathbf{p}(d + \beta a) + \boldsymbol{l}c - j_\sigma \boldsymbol{l}'b + \beta(\boldsymbol{l}b + j_\sigma \boldsymbol{l}'c) = \mathbf{p}(d + \beta a) + (\boldsymbol{l} + \beta \boldsymbol{l}')(c + \beta b),$$

where $\sigma = c + \beta b$, is real for each real value of d. So the lemma is proved.

Returning to the proof of the theorem: ∇ is transitive on the (ordered) quadrangles of Δ_0 and therefore certainly transitive on the incident pairs (R, r). Now let $\mathbf{h}, \boldsymbol{\rho}$ and $\mathbf{k}, \boldsymbol{\sigma}$ be any two point pairs R, C such that $\langle \mathbf{h}, \boldsymbol{\rho}\rangle$ and $\langle \mathbf{k}, \boldsymbol{\sigma}\rangle$ are complex lines. Write $\boldsymbol{\rho} = \mathbf{r}+\beta\mathbf{r}'$ and $\boldsymbol{\sigma} = \mathbf{s}+\beta\mathbf{s}'$. By the lemma, $\mathbf{H} = [\mathbf{h}, \mathbf{r}, \mathbf{r}']$ and $\mathbf{K} = [\mathbf{k}, \mathbf{s}, \mathbf{s}']$ are non-singular real matrices and we have a collineation in ∇ with matrix $\mathbf{KH}^{-1}$ such that $\mathbf{KH}^{-1}\mathbf{h} = \mathbf{k}$ and $\mathbf{KH}^{-1}\boldsymbol{\rho} = \boldsymbol{\sigma}$. It follows, in particular, that ∇ is transitive on the incident pairs (R, c); and, since every complex line contains a real point, that ∇ is transitive on the incident pairs (C, c).

The only case left is that of incident pairs (C, r). This case is easy, since we know that ∇ is transitive on complex points and that each complex point lies on exactly one real line.

Besides the projectivities $\mathscr{M}$, there are collineations derivable from the automorphisms of $\mathscr{Q}$. For, if $\mathscr{S} \in \mathrm{Aut}(\mathscr{Q})$ and $\boldsymbol{\lambda} = (\lambda_1, \lambda_2, \lambda_3)$ is a point, then $\mathscr{A}\boldsymbol{\lambda} = (\mathscr{S}\lambda_1, \mathscr{S}\lambda_2, \mathscr{S}\lambda_3)$ is a point and, if $\mathbf{p}$ is a real vector, $\mathscr{A}\mathbf{p} = \mathbf{p}$; thus

$$\begin{aligned}\langle \mathbf{p}, \boldsymbol{\lambda}\rangle &\to \{\mathscr{A}\mathbf{p}\} \cup \{\mathscr{A}(\mathbf{p}\kappa+\boldsymbol{\lambda}) : \kappa \in \mathscr{Q}\} \\ &= \{\mathbf{p}\} \cup \{\mathbf{p}(\mathscr{S}\kappa)+\mathscr{A}\boldsymbol{\lambda} : \kappa \in \mathscr{Q}\} \\ &= \langle \mathbf{p}, \mathscr{A}\boldsymbol{\lambda}\rangle.\end{aligned}$$

We have:

THEOREM 5.1.7 *Corresponding to the six automorphisms of $\mathscr{Q}$, there are six collineations*

$$\mathscr{A}_\sigma : (x_1, x_2, x_3) \to (\mathscr{S}x_1, \mathscr{S}x_2, \mathscr{S}x_3)$$

of Ψ (where $\mathscr{S}$ is the automorphism which sends α to $\sigma \in \mathscr{Q}^*$). Each collineation $\mathscr{A}_\sigma$ fixes all real points.

NOTATION 5.1.2 Σ: the collineation group consisting of the six collineations $\mathscr{A}_\sigma$.

EXERCISE 5.1.2 Show that $\mathscr{A} \in \Sigma$ and $\mathscr{M} \in \nabla \Rightarrow \mathscr{A}\mathscr{M} = \mathscr{M}\mathscr{A}$.

Finally we are to prove:

THEOREM 5.1.8 *Every collineation of Ψ can be expressed as $\mathscr{M}\mathscr{A}$, where $\mathscr{M} \in \nabla$ and $\mathscr{A} \in \Sigma$.*

To prove this theorem we need to anticipate a result proved later (Theorem 5.5.2), namely:

LEMMA 1 *Ψ has no non-trivial elation whose centre and axis are both complex.*

This implies that no collineation of Ψ maps a real point R to a complex point C. For, given any real line r through R, Ψ has non-trivial (R, r)-elations; and every collineation which maps R to C maps at least one of the four real lines through R to a complex line c through C (since there is only one real line through C). Thus:

LEMMA 2 *Every collineation of Ψ maps the real subplane Δ_0 onto itself.*

Now let $\mathscr{L}$ be any collineation of Ψ. From Lemma 2, there is a collineation $\mathscr{M} \in \nabla$ such that $\mathscr{M}\mathscr{L}$ fixes every point of Δ_0. Since $\mathscr{M}\mathscr{L}$ fixes the real line $x_3 = 0$ it maps the complex point $C = (\alpha, 1, 0)$ on this line to another complex point $C' = (\sigma, 1, 0)$ on it. So the collineation $\mathscr{A}_\sigma^{-1}\mathscr{M}\mathscr{L}$ fixes all real points and also the complex point C. By an argument used in the proof of Theorem 4.3.14, $\mathscr{A}_\sigma^{-1}\mathscr{M}\mathscr{L} = 1$; that is, $\mathscr{L} = \mathscr{M}^{-1}\mathscr{A}_\sigma$.

EXERCISE 5.1.3 Prove that the order of the full collineation group $\nabla\Sigma$ of Ψ is 33696.

EXERCISE 5.1.4 Show that Ψ is not a translation plane (i) with respect to any complex line (use Lemma 1 above); (ii) with respect to any real line (use Theorem 5.1.8).

5.2 Equations of lines and the self-duality of Ψ

In §5.1, we mentioned that the real lines of Ψ are given by the usual homogeneous linear equations for the lines of a plane of order 3; but we gave no equations for the complex lines.

Using (x, y, z) for coordinates, let us begin by choosing an ideal line $z = 0$. Consider a complex line $\langle \mathbf{p}, \boldsymbol{\lambda} \rangle$. If $\mathbf{p}$ does not lie on $z = 0$, we may write $\mathbf{p} = (r, s, 1)$, $\boldsymbol{\lambda} = (\lambda, \mu, \nu)$, and then

$$\langle \mathbf{p}, \boldsymbol{\lambda} \rangle = \langle \mathbf{p}, \boldsymbol{\lambda} - \mathbf{p}\nu \rangle,$$

where $\boldsymbol{\lambda} - \mathbf{p}\nu = (\lambda - r\nu, \mu - s\nu, 0)$. This point is complex, because the line is complex and $\mathbf{p}$ is the only real point on it, so that

$\lambda - r\nu \neq 0$, and we may write $\langle \mathbf{p}, \boldsymbol{\lambda} \rangle = \langle (r, s, 1), (1, \kappa, 0) \rangle$ for some $\kappa \in \mathscr{Q}^*$. It is now easy to verify that the nine proper points of the line are the solutions $(x, y, 1)$ of the equation

$$y - s = \kappa(x - r).$$

For clearly $(r, s, 1)$ satisfies this equation and, if $\theta \neq 0$, the point $(r, s, 1)\theta + (1, \kappa, 0) = ((r\theta + 1)\theta^{-1}, (s\theta + \kappa)\theta^{-1}, 1)$ satisfies

$$(y - s)\theta = (s\theta + \kappa) - s\theta = \kappa = \kappa(r\theta + 1 - r\theta) = \kappa(x - r)\theta;$$

that is, $y - s = \kappa(x - r)$. Thus any complex line $\langle \mathbf{p}, \boldsymbol{\lambda} \rangle$ with $\mathbf{p} \notin z = 0$ may be represented by such an equation, with $r, s \in \mathscr{D}$ and $\kappa \in \mathscr{Q}^*$.

Now take $\mathbf{p} \in z = 0$, say $\mathbf{p} = (r, s, 0)$; if the line is complex, then the second point by which the line is determined may be taken to be $(0, \kappa, 1)$, with $\kappa \in \mathscr{Q}^*$. The proper points $(x, y, 1)$ of the line satisfy the equation $ry = sx + r\kappa$. Summarizing, we have:

THEOREM 5.2.1 *The equations of the lines in Ψ are:*

(i) *Real lines* (13)

(i*a*) $y = mx + c \quad (m, c \in \mathscr{D})$, (9)

(i*b*) $x = c$, (3)

(i*c*) $z = 0$. (1)

(ii) *Complex lines* (78)

(ii*a*) $y - s = \kappa(x - r) \quad (r, s \in \mathscr{D}, \kappa \in \mathscr{Q}^*)$, (54)

(ii*b*) $y = mx + \kappa \quad (m \in \mathscr{D})$, (18)

(ii*c*) $x = \kappa$. (6)

There is a single form of homogeneous equation, due to G. Zappa (1957), which includes all the above, namely

$$ax + by + cz + \kappa(a'x + b'y + c'z) = 0,$$

where, for a complex line, $ax + by + cz = 0$ and $a'x + b'y + c'z = 0$ are any two of the four real lines through the real point on the line, and for a real line, $\kappa = 0$.

Returning to the equations of Theorem 5.2.1 above, we see that, because of the form of equation (ii*a*), we cannot ascribe any reasonable form for a coordinate-vector of a general line. With the exception of equation (ii$a$), the equations collectively, after multiplication on the right by z and replacement of xz, yz by x, y, are effectively symmetrical in (x, y, z), but in equation (ii$a$), modified to $y - sz = \kappa(x - rz)$, z behaves quite differently from x and y. That is, this formulation of the equations isolates a particular line $z = 0$ (the ideal line) as different from all the others. But in fact, since the projectivities of Δ_0 extend to projectivities in Ψ, $z = 0$ is projectively equivalent to any other real line; it is the method of constructing the equations which singles out $z = 0$ for special treatment.

EXERCISE 5.2.1 If $\boldsymbol{\lambda}, \boldsymbol{\rho}$ are complex points not on the same real line, show that the points of the set $\{\boldsymbol{\lambda}\} \cup \{\boldsymbol{\lambda}\kappa + \boldsymbol{\rho}\}$ are not collinear. (Use the equations of lines.)

EXERCISE 5.2.2 If $\mathbf{M}$ is a matrix with elements in $\mathscr{Q}$ such that $\mathbf{M}\theta$ is not real for any θ, show that $\mathbf{x} \to \mathbf{Mx}$ is not a collineation of Ψ.

EXERCISE 5.2.3 Prove that the equation of the line $\langle (r, s, 1), (\lambda, \mu, \nu) \rangle$ is

$$(\mu - s\nu)^{-1}(y - s) = (\lambda - r\nu)^{-1}(x - r),$$

provided $\mu \neq s\nu, \lambda \neq r\nu$.

THEOREM 5.2.2 *Ψ is self-dual.*

To prove this theorem we find two sets of equivalent parameters, one specifying a point and the other specifying a line, which are such that the condition of incidence between point and line is symmetrical in the two sets. We already have a set of parameters for a line (ii$a$), namely the ordered set $\langle r, s; \kappa \rangle$. In geometric terms, the parameters are:

$(r, s, 1)$: the coordinates of the single real point on the line,

$(1, \kappa, 0)$: the coordinates of the point in which the line meets $z = 0$.

Corresponding parameters for a point could be taken to be $\langle p, q; \mu \rangle$, where $px + qy + 1 = 0$ is the real line through the point, and $\mu x+y = 0$ is the join of the point to $(0,0,1)$. The point common to these two lines is $(1, -\mu, -p+q\mu)$.

The condition of incidence of the line $\langle r, s; \kappa\rangle$ and the point $\langle p, q; \mu\rangle$ is therefore

$$-\mu+s(p-q\mu) = \kappa(1+r(p-q\mu)).$$

Write $\nu = 1+rp-rq\mu$, then, if $qr \neq 0$, the equation becomes

$$pqrs-(1+sq)(1+rp-\nu) = \kappa\nu rq$$

or $$(1+sq-rq\kappa)\nu = (1+sq)(1+pr)-pqrs$$

or $$(1+sq-rq\kappa)(1+rp-rq\mu) = 1+pr+qs.$$

This equation is invariant under the interchanges

$$(r\leftrightarrow q, s\leftrightarrow p, \kappa\leftrightarrow\mu).$$

If $qr = 0$, then we take both $q = 0$, $r = 0$, and interchange $\langle 0, s; \kappa\rangle$ with $\langle p, 0; \mu\rangle$. Thus there is a 'partial correlation' $\mathscr{J}$ of the plane Ψ in which every line with equation of form (ii$a$)

$$y-s = \kappa(x-r)$$

is the correlative of the corresponding point $(1, -\kappa, -s+r\kappa)$.

To complete the definition of $\mathscr{J}$ we have to assign lines with other forms of equations to suitable points. Take first the real line (ia), $y = mx+c$. The point in which this line meets $z = 0$ is $(1, m, 0)$ and the other real points on the line are

$$(t, mt+c, 1) \quad (t\in\mathscr{D}).$$

In analogy with the notation already used, this line could be designated $\langle t, mt+c; m\rangle$, the correlative of the point with coordinates $(-1, m, (mt+c)-mt) = (-1, m, c)$; these are independent of the choice of t. That is, we select, for the correlative of the point $(-1, m, c)$, the line $y = mx+c$. A consistent system of correlatives can then be completed as shown in Table 5.2.1.

For the real points and lines we have

$$\mathscr{J}: (-1, m, c)\to[m, -1, c]\mathbf{x} = 0,$$

so that a pair of correlative real elements are pole and polar with regard to the conic $2xy+z^2 = 0$.

In §5.4 we prove that $\mathscr{J}$ is a polarity of Ψ and investigate further properties of this polarity.

Table 5.2.1 *Correlatives in* $\mathscr{J}$

(ii$a$)	
$\left.\begin{array}{r} y-s = \kappa(x-r) \\ \langle(r, s, 1), (1, \kappa, 0)\rangle \end{array}\right\} \leftarrow (-1, \kappa, s-r\kappa) \quad (\kappa \in \mathscr{Q}^*)$	
(ii$b$), (ia)	
$\left.\begin{array}{r} y = mx+\lambda \\ \langle(1, m, 0), (0, \lambda, 1)\rangle \end{array}\right\} \leftarrow (-1, m, \lambda) \quad (\lambda \in \mathscr{Q})$	
(ii$c$), (ib)	
$\left.\begin{array}{r} x = \lambda \\ \langle(0, 1, 0), (\lambda, 0, 1)\rangle \end{array}\right\} \leftarrow (0, -1, \lambda)$	
(ic)	
$\left.\begin{array}{r} z = 0 \\ \langle(1, 0, 0), (0, 1, 0)\rangle \end{array}\right\} \leftarrow (0, 0, 1)$	

5.3 Some constructions in Ψ

The following construction, which is a rearrangement of that in Diagram 3.1, plays a leading part in the main theorems in this section:

THEOREM 5.3.1 *If $XYZK$ is a regular quadrangle and $X, Y, Z \in \Delta_0$, then there is a unique set of three points A, B, C such that $A \in KX$ and $A \notin \{K, X\}$, $B \in KY$, $C \in KZ$ and $BC \supset X$, $CA \supset Y, AB \supset Z$.*

Take $\qquad K = (\lambda, \mu, 1) \;\; (\lambda, \mu \in \mathscr{Q} - \{0\}),$

and $\qquad X = (1, 0, 0), \quad Y = (0, 1, 0), \quad Z = (0, 0; 1).$

Since XK is $y = \mu$ and $A \in XK$ we may assume that for some ρ, $A = (\rho, \mu, 1)$ where $\rho \neq \lambda$, and then ZA is $y = \mu\rho^{-1}x$. YK is $x = \lambda$ and $Z \in AB$, so that $B = (\lambda, \mu\rho^{-1}\lambda, 1)$, XB is $y = \mu\rho^{-1}\lambda$, YA is $x = \rho$, and therefore $C = (\rho, \mu\rho^{-1}\lambda, 1)$. Thus

$$C \in ZK \Leftrightarrow \lambda^{-1}\rho = \mu^{-1}(\mu\rho^{-1}\lambda) \Leftrightarrow (\lambda^{-1}\rho)^2 = 1 \Leftrightarrow \rho = \pm\lambda$$

$$\Leftrightarrow \rho = -\lambda \quad \text{since} \quad \rho \neq \lambda.$$

We complete the diagram by naming the following points: $L = XK \cap YZ$, $M = YK \cap XZ$, $N = ZK \cap XY$, $L' = BC \cap YZ$, $M' = CA \cap XZ, N' = AB \cap XY$.

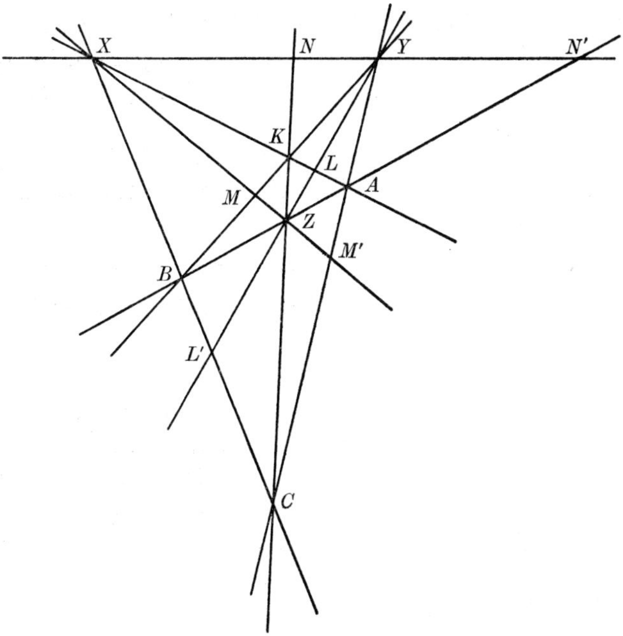

Diagram 5.3

The coordinates of the points in the figure are shown in Table 5.3.1.

Table 5.3.1

$X = (1, 0, 0)$,	$Y = (0, 1, 0)$,	$Z = (0, 0, 1)$	
$K = (\lambda, \mu, 1)$,	$A = (-\lambda, \mu, 1)$,	$B = (\lambda, -\mu, 1)$,	$C = (-\lambda, -\mu, 1)$
$L = (0, \mu, 1)$,	$M = (\lambda, 0, 1)$,	$N = (\lambda, \mu, 0)$	
$L' = (0, -\mu, 1)$,	$M' = (-\lambda, 0, 1)$,	$N' = (-\lambda, \mu, 0)$	

Table 5.3.2

0	1	2	3	4	5	6	7	8	9	10	11	12
X	Y	Z	.	L	K	B	.	.	.	L'	C	M'
Y	Z	N	.	K	B	A	.	.	.	C	M'	X
N	L	K	.	A	M	N'	.	.	.	X	Y	Z
N'	L'	C	.	X	Y	Z	.	.	.	B	A	M

For every selection of K the 13 points and the 9 collinear sets of four are distinct, and can be arranged as part of the cyclic pattern of a plane of order 3 as in Table 5.3.3 (compare Tables 3.1.1 and 3.1.2).

We are to prove that, according to the selection of K, the 13 points form either a subplane of order 3 or four overlapping subplanes of order 2.

THEOREM 5.3.2 (i) *If* $K \in \Delta_0$ *or* (ii) *if the real line though* K *contains* X, *or* Y *or* Z, *then the thirteen points in Table* 5.3.1 *form a subplane of order* 3.

(i) If K is real the points are all real and form Δ_0.

(ii) Let the real line through K be KZ, then we can select a coordinate system in Δ_0 such that the line is $x = y$, and by applying a suitable collineation of the group Σ we may transform K into the point $(\alpha, \alpha, 1)$. It is then easy to verify from Table 5.3.1, with $\lambda = \mu = \alpha$, that each of the four lines $y = \pm x \pm \alpha$ contains one of the sets of four points

$$MNL'A, \quad NLM'B, \quad LMN'C, \quad L'M'N'K.$$

THEOREM 5.3.3 *If* $K \notin \Delta_0$ *and the real line through* K *is distinct from* KX, KY, KZ, *then the thirteen points of Table* 5.3.1 *can be arranged as four overlapping Fano subplanes, namely*

$$XYZLMNK, \quad XYZLM'N'A, \quad XYZL'MN'B, \quad XYZL'M'NC.$$

We may suppose (using the collineations of ∇) that the real line through K is $y = x - 1$, and then (using a collineation of the group Σ) take K to be $(\alpha, \beta, 1)$. $L = (0, \beta, 1)$, $M = (\alpha, 0, 1)$, $N = (\alpha, \beta, 0)$, so that LM is $y - 1 = \gamma(x+1)$ and $N \in LM$; that is, the diagonal points of the quadrangle $XYZK$ are collinear. Similar arguments show that the other three sets are Fano subplanes.

The complete table of points and lines is shown in Table 5.3.3.

We now prove:

THEOREM 5.3.4 *Given in* Ψ *three non-collinear real points* X, Y, Z, *there are* 24 *complex points* F *such that the regular quadrangle* $XYZF$ *completes to a Fano subplane.*

Table 5.3.3

	Fano subplanes				Points				Collinear sets				Equations of lines
0	×	×	×	×	X	1	0	0	X	Y	N	N'	$[0, 0, 1]$
1	×	×	×	×	Y	0	1	0	Y	Z	L	L'	$x = 0$
2	×	×	×	×	Z	0	0	1	Z	N	K	C	$y = \gamma x$
3	×	×	.	.	N	α	β	0		N'	L'	M	$y+1 = -\gamma(x+1)$
4	×	.	×	.	L	0	β	1	L	K	A	X	$y = \beta$
5	×	.	.	.	K	α	β	1	K	B	M	Y	$x = \alpha$
6	.	.	.	×	B	α	$-\beta$	1	B	A	N'	Z	$y = -\gamma x$
7	.	.	×	.	A	$-\alpha$	β	1		M'	L	N'	$y-1 = -\gamma(x-1)$
8	×	.	.	×	M	α	0	1		M'	L'	N	$y+1 = \gamma(x-1)$
9	.	.	×	×	N'	$-\alpha$	β	0		N	M	L	$y-1 = \gamma(x+1)$
10	.	×	.	×	L'	0	$-\beta$	1	L'	C	X	B	$y = -\beta$
11	.	×	.	.	C	$-\alpha$	$-\beta$	1	C	M'	Y	A	$x = -\alpha$
12	.	×	×	.	M'	$-\alpha$	0	1	M'	X	Z	M	$y = 0$

F has to satisfy the condition that none of the lines FX, FY, FZ is real. The only real lines not through X or Y or Z are the four lines with equations $\pm x \pm y + z = 0$, so that F has to lie on one of these; on each of them there are six complex points. The coordinates of points of the set $\{F\}$ are $(\rho, \sigma, 1)$, $\rho, \sigma \in \mathscr{Q}^*$ and $\rho\sigma \notin \{\pm 1\}$—six choices for ρ followed by four for σ.

Let us now investigate the figure resulting from the DTP construction (Definition 2.6.1) which, in a plane $\Pi(\mathscr{K})$, generates the conic through a point K with tangents XZ at X and YZ at Y.

THEOREM 5.3.5 THE DTP CONSTRUCTION IN Ψ. *$XYZK$ is a regular quadrangle such that $X, Y, Z \in \Delta_0$. From these points Diagram 5.3 is constructed. If $\{R_i\}$ is the set of six points*

$$XA - \{X, A, K, L\}, \quad \textit{and} \quad S_i = YB \cap R_i Z, \quad T_i = R_i Y \cap S_i X,$$

then the set $\{T_i\}$ is the collinear set $AB - \{A, B, Z, N'\}$.

Let the coordinates of points be those set out in Table 5.3.1. $R_i \in XA$ and XA is $y = \mu$, so let R_i be $(\rho, \mu, 1)$ where

$$\lambda^{-1}\rho \notin \{0, \pm 1\} \quad \text{and} \quad R_i \neq X.$$

Similarly let S_i be $(\lambda, \sigma, 1)$. Then, since $Z \in R_i S_i$, $\lambda\sigma^{-1} = \rho\mu^{-1}$, XS_i is $y = \sigma$ and YR_i is $x = \rho$, so that

$$T_i = (\rho, \sigma, 1) = (\rho, \mu\rho^{-1}\lambda, 1).$$

Since

$$A = (-\lambda, \mu, 1) \quad \text{and} \quad B = (\lambda, -\mu, 1), \quad AB \text{ is } \lambda^{-1}x + \mu^{-1}y = 0,$$

and thus

$$T_i \in AB \Leftrightarrow \lambda^{-1}\rho + \rho^{-1}\lambda = 0 \Leftrightarrow (\lambda^{-1}\rho)^2 = -1$$
$$\Leftrightarrow \lambda^{-1}\rho \in \{\pm\alpha, \pm\beta, \pm\gamma\}.$$

That is, the set of points corresponding to

$$\{R_i\} = XA - \{X, A, K, L\} \quad \text{is} \quad \{T_i\} = AB - \{A, B, Z, N'\}.$$

The points of the set $\{T_i\}$ have coordinates $(\rho, \mu\rho^{-1}\lambda, 1)$; these are connected by the relation $y = \mu x^{-1}\lambda$ or $(\lambda^{-1}x)(\mu^{-1}y) = 1$. This equation is satisfied by the eight points $(\kappa, \mu\kappa^{-1}\lambda, 1)$, where

$\kappa \in \{\pm 1, \pm\alpha, \pm\beta, \pm\gamma\}$. The locus is completed by the 'ideal' points X, Y given as $(1, \mu\theta\lambda\theta, \theta)$ for $\theta = 0$ and $(\lambda\phi\mu\phi, 1, \phi)$ for $\phi = 0$.

The following exercise is concerned with the 'angle-in-a-semicircle' construction. We cannot achieve in Ψ quite the elegance of the corresponding construction in Ω (Exercise 4.6.6), because it is not possible to split the points of any line in Ψ into five pairs which are preserved under all collineations fixing the line. In any case there is no line of Ψ which, like the ideal line in Ω, is fixed by all collineations. We can, however, split the points of any real line, say XY, into a set of four pairs 'in an involution' together with the 'double points' X, Y of the 'involution', in such a way that every collineation which fixes $\{X, Y\}$ permutes the four pairs.

EXERCISE 5.3.1 $\{N_i, N_i'\}$ are the pairs

$$\{(\rho, 1, 0), (-\rho, 1, 0) : \rho \in \mathcal{Q}^*\} \quad \text{on} \quad XY, \quad \text{and} \quad XYDD'$$

is a regular quadrangle in Δ_0. Prove that the set of six points $\{P_i = DN_i \cap D'N_i'\}$ lies on one of the sides of the diagonal triangle of $XYDD'$.

5.4 Polarities in Ψ

In investigating (§5.2) the duality property of Ψ we found that there is a correlation $\mathcal{J}$ under which the line $y - s = \kappa(x - r)$ corresponds to the point $(-1, \kappa, s - r\kappa)$, and further that the real absolute points of $\mathcal{J}$ form the conic $xy - z^2 = 0$. Following the pattern of the Galois plane Φ, we could define, as an analogue in Ψ of a conic in Φ, the set of points which lie on their correlative lines in $\mathcal{J}$, and, as an analogue of the Hermitian set, the set of points which lie on the conjugates of their correlative lines in $\mathcal{J}$. The subset of real points in each of these sets is the conic $xy - z^2 = 0$.

Under $\mathcal{J}$ we have, for complex lines and points, the correspondences

$$\begin{aligned}
y - s = \kappa(x - r) &\leftarrow (-1, \kappa, -r\kappa + s) && (\kappa \in \mathcal{Q}^*),\\
y = mx + \kappa &\leftarrow (-1, m, \kappa) && (m \in \mathcal{D}),\\
x = \mu \quad &\leftarrow (0, -1, \mu) && (\mu \in \mathcal{Q}^*),
\end{aligned}$$

so that the complex points in the set ψ of absolute points are the solutions of one or other of the equations

$$\kappa + s(r\kappa - s) = \kappa(-1 + r(r\kappa - s)), \tag{i}$$

$$m = -m + \kappa^2 \tag{ii}$$

(for no value of μ does the point $(0, -1, \mu)$ lie on $x = \mu$).

Equation (ii), which is equivalent to $\kappa^2 = -m$, has six solutions: $m = 1, \kappa \in \{\pm\alpha, \pm\beta, \pm\gamma\}$. That is, from equation (ii) we derive six points of ψ, namely $(-1, 1, \kappa)$; these all lie on the line $x + y = 0$.

Equation (i) may be written as

$$(1 + rs)\kappa - s^2 = \kappa(-1 - rs + r^2\kappa) \tag{iii}$$

and we can classify the solutions of this equation according to the three possible values of $1 + rs$.

$$1 + rs = 0 \Rightarrow r^2 = s^2 = 1 \Rightarrow \kappa^2 = -1,$$

so that there are two sets of six solutions, namely

$$r = 1, \quad s = -1, \quad \kappa \in \mathscr{Q}^*,$$

giving six points $(-1, \kappa, -1 - \kappa)$ lying on $-x + y + 1 = 0$, and

$$r = -1, \quad s = 1, \quad \kappa \in \mathscr{Q}^*,$$

giving six points $(-1, \kappa, 1 + \kappa)$ lying on $x - y + 1 = 0$.

$$1 + rs = 1 \Rightarrow \left\{ \begin{matrix} r^2 = 1, s = 0 \\ \text{or } r = 0, s^2 = 1 \\ \text{or } r = s = 0 \end{matrix} \right\} \Rightarrow \kappa = 0 \quad \text{or} \quad -1$$

and

$$1 + rs = -1 \Rightarrow r^2 = s^2 = 1 \Rightarrow -\kappa - 1 = \kappa(1 + \kappa) \Rightarrow \kappa = -1;$$

so that, since $\kappa \in \mathscr{Q}^*$, there is no solution in either case.

THEOREM 5.4.1 *The set* Ψ *of absolute points in the correlation* $\mathscr{J}$ *consists of 22 points, namely the real quadrangle* $(1, 0, 0)$, $(0, 1, 0)$, $(1, 1, 1)$, $(-1, -1, 1)$ *and the sets of six complex points on the sides of the diagonal triangle of this quadrangle.*

It may be remarked that the quadrangle together with any one of the collinear sets forms the DTP configuration of Theorem 5.3.5.

THEOREM 5.4.2 *$\mathscr{J}$ is a polarity; that is, $Q \in \mathscr{J}P \Rightarrow P \in \mathscr{J}Q$.*

Assume first that both lines $\mathscr{J}P$ and $\mathscr{J}Q$ are given by equations of type (ii$a$), say

$$P = (-1, \kappa, s - r\kappa), \quad \mathscr{J}P \quad \text{is} \quad y - s = \kappa(x - r),$$

$$Q = (-1, \lambda, v - u\lambda), \quad \mathscr{J}Q \quad \text{is} \quad y - v = \lambda(x - u).$$

The two conditions are

$$Q \in \mathscr{J}P\colon (1 + su)\lambda - sv = \kappa(-1 - rv + ru\lambda),$$

$$P \in \mathscr{J}Q\colon (1 + rv)\kappa - sv = \lambda(-1 - su + ru\kappa).$$

First, if $ru = 0$, these two relations are identical, so that if the real point on either $\mathscr{J}P$ or $\mathscr{J}Q$ lies on $x = 0$ the condition $Q \in \mathscr{J}P \Rightarrow P \in \mathscr{J}Q$ is satisfied. Assume therefore $ru \neq 0$, and write

$$\mu = -1 - rv + ru\lambda, \quad \nu = -1 - su + ru\kappa,$$

so that the conditions become

$$Q \in \mathscr{J}P\colon (1 + su)(1 + rv + \mu) - rsuv = (1 + su + \nu)\mu,$$

$$P \in \mathscr{J}Q\colon (1 + rv)(1 + su + \nu) - rsuv = (1 + rv + \mu)\nu.$$

That is, $\quad 1 + su + rv = \nu\mu \quad$ and $\quad 1 + su + rv = \mu\nu.$

In each of these relations the left-hand member is real, so that $\mu\nu$ is real; that is, $\mu = \pm\nu$, and therefore $\mu\nu = \nu\mu$. Thus

$$Q \in \mathscr{J}P \Rightarrow \mu\nu = \pm 1 = 1 + su + rv \Rightarrow P \in \mathscr{J}Q.$$

EXERCISE 5.4.1 Verify that $Q \in \mathscr{J}P \Rightarrow P \in \mathscr{J}Q$ for other forms of equations of the lines $\mathscr{J}P$ and $\mathscr{J}Q$.

EXERCISE 5.4.2 Find the correlative line of the point $(-1, \kappa, \lambda)$, where $\kappa = k + \beta k'$, $\lambda = l + \beta l'$, and in terms of k, k', l, l' the condition that the point lies on its correlative line.

From the point of view of configurational geometry, the analogue in Ψ of the Hermitian set in Φ (§ 3.8) has greater interest than ψ. Moreover, in $\mathscr{Q}$, the analogue of the conjugacy operation in $\mathscr{F}$, namely the operation $\mathscr{B}: \alpha \leftrightarrow -\gamma$, $\beta \leftrightarrow -\beta$, $\gamma \leftrightarrow -\alpha$, is one of three equivalent operations $\mathscr{A}$, $\mathscr{B}$, $\mathscr{C}$ (§ 1.4), so that in Ψ there are three 'Hermitian sets' associated with the conic-analogue ψ.

Write, for each element κ of $\mathscr{Q}$, $\kappa^* = \mathscr{B}\kappa$, and take $\mathscr{J}^*$ to be the correlation

$$(-1, \kappa, s-r\kappa) \to y-s = \kappa^*(x-r)$$

$$(-1, m, \kappa) \to \quad y = mx+\kappa^*,$$

$$(0, -1, \mu) \to \quad x = \mu^*,$$

$$(0, 0, 1) \to \quad z = 0.$$

EXERCISE 5.4.3 Show that $\mathscr{J}^*$ is a polarity.

Now consider the set of absolute points $\psi^* = \{P: P \in \mathscr{J}^*P\}$. For real points, $P = P^*$, so that the real points of the set ψ^* are the points of the conic $xy - z^2 = 0$. The complex points satisfy one of the three conditions:

$$(-1, \kappa, s-r\kappa) \in y-s = \kappa^*(x-r), \tag{i}$$

or

$$(-1, m, \kappa) \in \quad y = mx+\kappa^*, \tag{ii}$$

or

$$(0, -1, \mu) \in \quad x = \mu^*. \tag{iii}$$

Again no point can satisfy condition (iii). Condition (ii) is $-m = \kappa\kappa^*$, the only solutions of which are $m = -1$, $\kappa = \pm\beta$, so that from condition (ii) we derive the pair of points $(1, 1, \pm\beta)$, of which the correlative lines are $y = -x \pm \beta$.

From condition (i) we have

$$(1+rs)\kappa - s^2 = \kappa^*(-1-rs+r^2\kappa).$$

If $r = 0$, $\kappa - s^2 = -\kappa^*$ of which the solutions are

$$\kappa = -\alpha, \quad \kappa^* = \gamma, \quad s = \pm 1, \quad P = (-1, -\alpha, \pm 1),$$

$$\kappa = \pm\beta, \quad \kappa^* = \mp\beta, \quad s = 0, \quad P = (-1, \pm\beta, 0),$$

$$\kappa = \gamma, \quad \kappa^* = -\alpha, \quad s = \pm 1, \quad P = (-1, \gamma, \pm 1).$$

If $r \neq 0$, so that $r^2 = 1$, write $\nu = -1 - rs + \kappa$, then

$$(1+rs)(\nu+1+rs) - s^2 = (\nu^* + 1 + rs)\nu;$$

that is, $$1 - rs = \nu^*\nu.$$

$\nu^*\nu$ is real only when $\nu = \pm\beta$, $\nu^*\nu = 1$, so that solutions are obtained only when $s = 0$ (since $r \neq 0$) and $\kappa = \nu + 1$. The solutions are

$$r = \pm 1, \quad s = 0, \quad \nu = \beta, \quad \kappa = \alpha, \quad P = (-1, \alpha, \mp\alpha),$$

$$r = \pm 1, \quad s = 0, \quad \nu = -\beta, \quad \kappa = -\gamma, \quad P = (-1, -\gamma, \pm\gamma).$$

Summarizing the results we have:

THEOREM 5.4.3 *The Hermitian-set analogue ψ^* consists of four real points on the conic $xy = z^2$, together with twelve complex points; the sixteen points may be arranged as four regular quadrangles $\mathfrak{R}$, $\mathfrak{L}$, $\mathfrak{M}$, $\mathfrak{N}$, namely*

$\mathfrak{R}$	$\mathfrak{M}$	$\mathfrak{L}$	$\mathfrak{N}$
$1, 0, 0$	$1, \alpha, 1$	$\beta, \beta, 1$	$\alpha, 1, 1$
$0, 1, 0$	$1, -\gamma, 1$	$-\beta, -\beta, 1$	$-\gamma, 1, 1$
$1, 1, 1$	$-\alpha, -1, 1$	$1, \beta, 0$	$-1, -\alpha, 1$
$-1, -1, 1$	$\gamma, -1, 1$	$1, -\beta, 0$	$-1, \gamma, 1.$

By constructing Table 5.4.1 we find the following incidence relations among the sixteen points.

THEOREM 5.4.4 *The diagonal triangle of each of the quadrangles, $\mathfrak{R}$, $\mathfrak{L}$, $\mathfrak{M}$, $\mathfrak{N}$ is $\{(1, 1, 0), (-1, 1, 1), (1, -1, 1)\}$, and each pair of quadrangles has one pair of opposite sides common.*

That is, through each vertex of the diagonal triangle there pass four lines, each containing two pairs of the sixteen points, one pair from each of two of the quadrangles.

In Table 5.4.1 the four vertices of each quadrangle have been assigned the indices 1, 2, 3, 4 in such a way that the twelve collinear sets are:

$(1, 1, 0)$: $\mathfrak{R}_1\mathfrak{R}_4\mathfrak{L}_2\mathfrak{L}_3$, $\mathfrak{R}_2\mathfrak{R}_3\mathfrak{L}_1\mathfrak{L}_4$, $\mathfrak{M}_1\mathfrak{M}_4\mathfrak{N}_1\mathfrak{N}_4$, $\mathfrak{M}_2\mathfrak{M}_3\mathfrak{N}_2\mathfrak{N}_3$,

$(1, -1, 1)$: $\mathfrak{R}_2\mathfrak{R}_4\mathfrak{M}_1\mathfrak{M}_3$, $\mathfrak{R}_1\mathfrak{R}_3\mathfrak{M}_2\mathfrak{M}_4$, $\mathfrak{L}_2\mathfrak{L}_4\mathfrak{N}_2\mathfrak{N}_4$, $\mathfrak{L}_1\mathfrak{L}_3\mathfrak{N}_1\mathfrak{N}_3$,

$(-1, 1, 1)$: $\mathfrak{R}_3\mathfrak{R}_4\mathfrak{N}_1\mathfrak{N}_2$, $\mathfrak{R}_1\mathfrak{R}_2\mathfrak{N}_3\mathfrak{N}_4$, $\mathfrak{L}_3\mathfrak{L}_4\mathfrak{M}_3\mathfrak{M}_4$, $\mathfrak{L}_1\mathfrak{L}_2\mathfrak{M}_1\mathfrak{M}_2$.

Table 5.4.1 *The Hermitian-set analogue,* ψ^*

Vertex (1, 1, 0)

Lines	$z = 0$		$x = y$		$y = x+\beta$		$y = x-\beta$	
Set $\mathfrak{K}$	1, 0, 0	1	1, 1, 1	2	.	.	.	.
	0, 1, 0	4	$-1, -1, 1$	3	.	.	.	.
Set $\mathfrak{L}$	$1, \beta, 0$	3	$\beta, \beta, 1$	1	.	.	.	.
	$1, -\beta, 0$	2	$-\beta, -\beta, 1$	4	.	.	.	.
Set $\mathfrak{M}$	.	.	.	.	$1, \alpha, 1$	1	$1, -\gamma, 1$	3
	.	.	.	.	$-\alpha, -1, 1$	4	$\gamma, -1, 1$	2
Set $\mathfrak{N}$	.	.	.	.	$-1, \gamma, 1$	1	$-1, -\alpha, 1$	2
	.	.	.	.	$-\gamma, 1, 1$	4	$\alpha, 1, 1$	3

Vertex (1, −1, 1)

Lines	$x = 1$		$y = -1$		$y+1 = \beta(x-1)$		$y+1 = -\beta(x-1)$	
Set $\mathfrak{K}$	1, 1, 1	2	1, 0, 0	1	.	.	.	.
	0, 1, 0	4	$-1, -1, 1$	3	.	.	.	.
Set $\mathfrak{L}$	.	.	.	.	$1, \beta, 0$	3	$1, -\beta, 0$	2
	.	.	.	.	$\beta, \beta, 1$	1	$-\beta, -\beta, 1$	4
Set $\mathfrak{M}$	$1, -\gamma, 1$	3	$\gamma, -1, 1$	2	.	.	.	.
	$1, \alpha, 1$	1	$-\alpha, -1, 1$	4	.	.	.	.
Set $\mathfrak{N}$	.	.	.	.	$\alpha, 1, 1$	3	$-\gamma, 1, 1$	4
	.	.	.	.	$-1, \gamma, 1$	1	$-1, -\alpha, 1$	2

Vertex (−1, 1, 1)

Lines	$x = -1$		$y = 1$		$y-1 = -\beta(x+1)$		$y-1 = \beta(x+1)$	
Set $\mathfrak{K}$	$-1, -1, 1$	3	1, 0, 0	1	.	.	.	.
	0, 1, 0	4	1, 1, 1	2	.	.	.	.
Set $\mathfrak{L}$	.	.	.	.	$1, -\beta, 0$	2	$1, \beta, 0$	3
	.	.	.	.	$\beta, \beta, 1$	1	$-\beta, -\beta, 1$	4
Set $\mathfrak{M}$	.	.	.	.	$1, \alpha, 1$	1	$1, -\gamma, 1$	3
	.	.	.	.	$\gamma, -1, 1$	2	$-\alpha, -1, 1$	4
Set $\mathfrak{N}$	$-1, -\alpha, 1$	2	$\alpha, 1, 1$	3	.	.	.	.
	$-1, \gamma, 1$	1	$-\gamma, 1, 1$	4	.	.	.	.

The three sets ψ^* which correspond to the one set ψ have the same real quadrangle. The three sets of three complex quadrangles contain among them all thirty-six complex points on the six sides of the real quadrangle.

5.5 A combinatorial analysis of Ψ

The plane Ψ has a projectivity $\mathscr{G} \in \nabla$ corresponding to the Singer matrix **G**, which played so significant a part in Chapter 3. For convenience of reference we reproduce in Table 5.5.1 the table of powers of **G**.

Table 5.5.1 = 3.2.1

The columns forming $\mathbf{G}^r$ are those numbered r, $r+1$, $r+2$.

0	1	2	3	4	5	6	7	8	9	10	11	12
1	0	0	1	0	1	1	1	−1	−1	0	1	−1
0	1	0	1	1	1	−1	−1	0	1	−1	1	0
0	0	1	0	1	1	1	−1	−1	0	1	−1	1

From any point P of Ψ we derive thirteen points $\mathscr{G}^r P$, so that $\mathscr{G}$ separates Ψ into seven disjoint sets of thirteen points. One of these forms the real subplane, Δ_0, but the others do not form subplanes of order three. They do, however, provide a very serviceable method of naming the points which enables us more easily than by direct algebraic methods to investigate combinatorial properties of the plane.

Table 5.5.2 *Designations of complex points and lines of* Ψ

Points on $z = 0$		Lines through X		Other points and lines
$\alpha, 1, 0$	A_0	$\alpha y = 1$	a_0	$\{U_r\} = \{A_r, B_r, C_r, A'_r, B'_r, C'_r\}$
$\beta, 1, 0$	B_0	$\beta y = 1$	b_0	$\{u_r\} = \{a_r, b_r, c_r, a'_r, b'_r, c'_r\}$
$\gamma, 1, 0$	C_0	$\gamma y = 1$	c_0	$U_r = \mathscr{G}^r U_0$
$-\gamma, 1, 0$	A'_0	$-\gamma y = 1$	a'_0	.
$-\beta, 1, 0$	B'	$-\beta y = 1$	b'_0	$u_r = \mathscr{G}^r u_0$
$-\alpha, 1, 0$	C'_0	$-\alpha y = 1$	c'_0	.

In this notation we have, since Aut $(\mathscr{Q})$ is generated by $\mathscr{A}$ and $\mathscr{C}$ (Theorem 1.4.1):

THEOREM 5.5.1 *The collineation group* Σ *may be generated by the two collineations*

$$\mathscr{A}: (K_r, A_r, B_r, C_r, A_r', B_r', C_r') \to (K_r, C_r', A_r', B_r', B_r, C_r, A_r),$$

$$\mathscr{C}\mathscr{A}: (K_r, A_r, B_r, C_r, A_r', B_r', C_r') \to (K_r, B_r, C_r, A_r, C_r', A_r', B_r').$$

The transforms of the lines u_r form identical patterns.

We now construct the analogue of Table 3.5.1, using the miniquaternion elements of $\mathscr{Q}$ in place of the Galois elements of $\mathscr{F}$, and provide names for the lines $U_r X$.

Table 5.5.3

r	$A_r = (x, y, z)$			$z \in \{\pm 1, \pm\beta\}$	zy^{-1}	$A_r X$		
0	α	1	0	.	0	k_0	.	.
1	0	α	1	✓	$-\alpha$	.	.	c_0'
2	1	1	α	.	α	.	a_0	.
3	α	γ	1	✓	$-\gamma$	.	.	a_0'
4	1	γ	γ	.	1	k_4	.	.
5	γ	β	γ	.	α	.	a_0	.
6	γ	$-\gamma$	β	✓	α	.	a_0	.
7	β	$-\alpha$	$-\gamma$	.	$-\beta$	.	.	b_0'
8	$-\gamma$	1	$-\alpha$	.	$-\alpha$	.	.	c_0'
9	$-\alpha$	β	1	✓	$-\beta$	.	.	b_0'
10	1	$-\beta$	β	✓	-1	k_{10}	.	.
11	β	α	$-\beta$	✓	$-\gamma$	.	.	a_0'
12	$-\beta$	0	α	.	.	k_{12}	.	.

From this table we see that the points A_r and the lines $A_r X$ have the following incidences ($u_0 \supset A_r$):

u_0	a_0	b_0	c_0	a_0'	b_0'	c_0'
A_r: r	2, 6, 5	.	.	3, 11	9, 7	1, 8

For $r \in \{0, 4, 10, 12\}$ $A_r X$ is the real line k_r. The corresponding tables for the other five members of the set $\{u_r\}$ are derived from Table 5.5.3 by the substitutions listed in Theorem 5.5.1; the resulting complete set of tables of incidence is:

Table 5.5.4

(a) $k_r = \{K_r, K_{r+1}, K_{r+3}, K_{r+9}\} \cup \{U_r\}$,
$K_r = k_r \cap k_{r-1} \cap k_{r-3} \cap k_{r-9} \cap a_r \cap \ldots \cap c'_r$.

(b) Sets of points constituting the lines u_0:

Line	K_r	A_r	B_r	C_r	A'_r	B'_r	C'_r
a_0	0	2, 6, 5	.	.	3, 11	9, 7	1, 8
b_0	0	.	2, 6, 5	.	9, 7	1, 8	3, 11
c_0	0	.	.	2, 6, 5	1, 8	3, 11	9, 7
a'_0	0	3, 11	9, 7	1, 8	2, 6, 5	.	.
b'_0	0	9, 7	1, 8	3, 11	.	2, 6, 5	.
c'_0	0	1, 8	3, 11	9, 7	.	.	2, 6, 5

(c) $U_r \in v_0 \Rightarrow U_{r+s} \in v_s, \quad r, s \in \{0, \ldots, 12\}$.

These are tables of incidences in a projective plane so that the line joining two given points has to be uniquely determinable; many examples of determinations will be found in Theorem 5.5.2. In particular, because of the relations (c), table (b) must have the property that the differences between elements in two given rows which lie in the same columns will form the set $\{1, \ldots, 12\}$. For example:

'$b_0 - c_0$': (8, 6, 1, 12) (11, 5, 3, 10) (7, 2, 9, 4)
'$a_0 - c'_0$': (1, 5, 4, 7, 11, 10) (12, 8, 9, 6, 2, 3)

The equations of the line u_r can be obtained by applying the transformation $\mathscr{G}^r$ to u_0. Thus, write the equation of a_0 as $y = -\alpha$; that is, as

$$a_0\colon\ [0, 1, 0]\,\mathbf{x} = -\alpha[0, 0, 1]\,\mathbf{x},$$

then, since $a_r = \mathscr{G}^r a_0$, we have for a_r,

$$a_r\colon\ [0, 1, 0]\,\mathbf{G}^{-r}\mathbf{x} = -\alpha[0, 0, 1]\,\mathbf{G}^{-r}\mathbf{x}$$

or
$$\mathbf{m}_r^T\mathbf{x} = -\alpha\mathbf{n}_r^T\mathbf{x},$$

where $\mathbf{m}_r^T$ and $\mathbf{n}_r^T$ are the second and third rows of $\mathbf{G}^{-r}$. For example, the equations of a_8 is:

$$a_8\colon\ [1, -1, -1]\,\mathbf{x} = -\alpha[1, 1, -1]\,\mathbf{x}$$

or
$$x - y - 1 = -\alpha(x + y - 1)$$

or
$$(x - y - 1) + (x + y - 1) = (-\alpha + 1)(x + y - 1)$$

or
$$-x+1 = -\beta(x+y-1)$$
or
$$-\beta(x-1) = x-1+y$$
or
$$y = -(\beta+1)(x-1) = -\alpha(x-1).$$

The complete set of equations of the lines a_r is listed in Table 5.5.5.

Table 5.5.5 *Equations of lines a_r*

a_0	$y = -\alpha$	a_7	$y-1 = -\alpha(x+1)$
a_1	$0 = x-\alpha$	a_8	$y = -\alpha(x-1)$
a_2	$y = \gamma x$	a_9	$y = -x+\alpha$
a_3	$y = x-\gamma$	a_{10}	$y+1 = \gamma x$
a_4	$y-1 = \beta x$	a_{11}	$y+1 = -\alpha(x+1)$
a_5	$y-1 = \alpha(x-1)$	a_{12}	$y = \alpha(x+1)$
a_6	$y+1 = \beta(x-1)$		

The equations of the lines $b_r, c_r, \ldots, c'_r$ are derived directly from these by operations of the group Aut $\mathscr{Q}$ on the one complex constant which appears in each of these equations. It should be remembered also that if the equation of u_r is $y-b = \kappa(x-a)$ then $(a, b, 1)$ is the real point on u_r, namely K_r, and the line is the set of points $\{(a, b, 1)\,\theta+(1, \kappa, 0)\}$.

EXERCISE 5.5.1 Name (in the $A, B, C, \ldots$ notation) all the points and lines in the configuration in Diagram 5.3, when $X = K_0$, $Y = K_1$, $Z = K_2$ and (i) $K = A_2$, (ii) $K = A_3$. Verify that the configurations have the properties stated in Theorems 5.3.2 and 5.3.3.

We conclude this section by using Table 5.5.4 to prove the crucial Lemma 1 in the proof of Theorem 5.1.8, namely:

THEOREM 5.5.2 *There is no non-trivial elation in Ψ which has a complex point as centre and a complex line as axis.*

Since the group of projectivities ∇ is transitive on incident (complex point)–(complex line) pairs, it is sufficient to prove that there is no elation for any one such incident pair (D, d). Further, it will be sufficient to select any point P disjoint from D and d and show that there is no elation which maps P to U for each selection of the point U from the eight points of DP other than D and P. Let Q, R be any two points disjoint from DP and d and not collinear with P, nor with D, and construct, using a selected point U, the points L, M, N, V, W, where

$$d \cap (PQ, PR, QR) = (N, M, L),$$

$$UN \cap DQ = V, \quad UM \cap DR = W.$$

Then there is an elation with centre D and axis d which maps P to U only if $VW \supset L$ (§ 2.2).

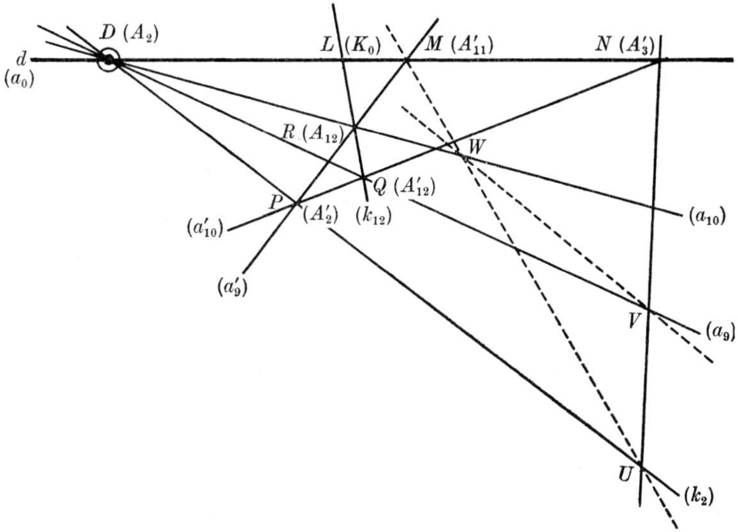

Diagram 5.5

Take:

$$D = A_2, \quad d = a_0, \quad P = A_2', \quad Q = A_{12}', \quad R = A_{12},$$

so that $DP = k_2$, $DQ = a_9$, $DR = a_{10}$, $PQ = a_{10}'$, $PR = a_9'$, $QR = k_{12}$, $N = A_3'$, $M = A_{11}'$, $L = K_0$, and

$$U \in \{K_2, K_3, K_5, K_{11}, B_2, B_2', C_2, C_2'\}.$$

The condition that has to be satisfied is $K_0 \in VW$; that is,

$$VW \in \{k_0, k_{10}, k_4, b_0, c_0, a_0', b_0', c_0'\}.$$

Table 5.5.6

U	UA'_3	$UA'_3 \cap a_9 = V$	UA'_{11}	$UA'_{11} \cap a_{10} = W$	VW	$VW \cap a_0$	$VW \supset K_0$?
K_2	c_2	B'_5	b_2	C'_5	k_5	A_5	No
K_3	k_3	B'_3	c_3	B'_6	b'_1	B'_7	No
K_5	a_5	A_{11}	a'_5	A_3	a'_0	K_0	Yes
K_{11}	a'_{11}	A_1	k_{11}	C'_{11}	c'_6	C'_8	No
B_2	b_9	K_9	a'_6	A'_8	k_8	C'_8	No
B'_2	b_7	C'_{10}	a_8	B'_4	c_1	C'_8	No
C_2	a'_1	A'_7	c_{10}	K_{10}	k_7	B'_7	No
C'_2	c_8	C'_4	b_4	A'_0	b_6	B'_7	No

From the table we see that the only possible (A_2, a_0)-elation would be one mapping A'_2 to K_5. But this elation has also to map K_2 to some point of k_2. In the diagram take $P = K_2$, $Q = B'_5, R = C'_5$, so that $QR \cap a_0 = k_5 \cap a_0 = A_5$. From any point U, $U \notin \{A_2, A'_2, K_2, K_5\}$, of k_2 construct V and W. For every such selection of U, $VW \not\ni A_5$, that is $VW \cap QR \notin a_0$, so that, since no image point for K_2 can be found, there is no (A_2, a_0)-elation mapping A'_2 to K_5, and therefore no (A_2, a_0)-elation at all.

EXERCISE 5.5.2 Verify that if $D = A_2$, $d = a_0$, $P = K_2$, $Q = K_9$, $R = K_1$ the required incidence condition is satisfied for every selection of the point U. Is this relevant to the main theorem?

5.6 The derivation of Ψ from Φ

We are now to examine in greater detail the relation between the Φ-table 3.5.2 and the Ψ-table 5.5.4. Combined to display the relation most clearly, the first three columns of the two tables form Table 5.6.1.

We observe first that in any column the set of indices in a pair of conjugate rows u_0, u'_0 is the same for Ψ as for Φ. In each Φ-column five of the entries have been starred; the effect of transferring each of these to the conjugate row is to produce the corresponding Ψ-column. To see clearly the relation between the two systems we need first Table 5.6.2 of equivalences (under addition) of the elements of $\mathscr{F} - \mathscr{D}$ and $\mathscr{Q}^*$.

Table 5.6.1

	A		B		C	
	Φ	Ψ	Φ	Ψ	Φ	Ψ
a_0	2, 3*, 5, 11*	2, 6, 5	9*	.	1*	.
b_0	9*	.	5, 6, 8*, 1*	2, 6, 5	3*	.
c_0	1*	.	3*	.	6, 7*, 9*, 2	2, 6, 5
a_0'	6*	3, 11	7	9, 7	8	1, 8
b_0'	7	9, 7	2*	1, 8	11	3, 11
c_0'	8	1, 8	11	3, 11	5*	9, 7

Table 5.6.2

$\mathscr{F}-\mathscr{D}$	ω	$\omega-1$	$\omega+1$	$-\omega$	$-\omega+1$	$-\omega-1$
	ω	$-\omega^2$	$-\omega^3$	$-\omega = -\omega^{3*}$	$\omega^2 = -\omega^{2*}$	$\omega^3 = \omega^*$
$\mathscr{Q}^*$	α	$\alpha-1$	$\alpha+1$	$-\alpha$	$-\alpha+1$	$-\alpha-1$
	α	β	γ	$-\alpha = \gamma^*$	$-\beta = \beta^*$	$-\gamma = \alpha^*$

In Table 5.6.3 the various columns are:

I The list of coordinate vectors $\mathbf{G}^r(\omega, 1, 0), r = 0, \ldots, 12$; that is, the coordinate-vectors of the points $A_0, \ldots, A_{12}$ in the subplane Δ_A (copied from Table 3.5.1).

II The lines u_0 (or k_t) in Φ on which the points A_r lie.

III The lines in Ψ on which the points A_r lie, from Table 5.5.4.

IV The list of vectors $\mathbf{G}^r(\alpha, 1, 0)$.

V The list of vectors in column IV with each vector (ρ, σ, τ), $\tau \neq 0$, replaced by $(\rho\tau^{-1}, \sigma\tau^{-1}, 1)$, under miniquaternion multiplication.

Let us now star the vectors in column I corresponding to pairs of symbols in columns II, III one of which is u_0 and the other u_0'. That is, we star the vectors in rows 1, 3, 6, 9, 11. The third component of each of these vectors is an even power of ω, namely ± 1 or $\pm\omega^2$. The only other vector in the list with this property is that in row 10; in both Φ and Ψ, A_{10} lies on the real line k_{10}. That is, an entry in column I, which corresponds in Φ to a point A_r on u_0, corresponds in Ψ to a point A_r on u_0' if and only

Table 5.6.3 *Subsets A_r in Φ and Ψ*

r	I			II(Φ)			III (Ψ)			IV			V		
0	ω	1	0	k_0	.	.	k_0	.	.	α	1	0	α	1	0
1	0	ω	1*	.	.	c_0	.	.	c_0'	0	α	1	0	α	1
2	1	1	ω	.	a_0	.	.	a_0	.	1	1	α	$-\alpha$	$-\alpha$	1
3	ω	$-\omega^3$	1*	.	a_0	.	.	a_0'	.	α	γ	1	α	γ	1
4	1	$-\omega^3$	$-\omega^3$	k_4	.	.	k_4	.	.	1	γ	γ	$-\gamma$	1	1
5	$-\omega^3$	$-\omega^2$	$-\omega^3$	.	a_0	.	.	a_0	.	γ	β	γ	1	$-\alpha$	1
6	$-\omega^3$	ω^3	$-\omega^2$*	.	a_0'	.	.	a_0	.	γ	$-\gamma$	β	α	$-\alpha$	1
7	$-\omega^2$	$-\omega$	ω^3	.	.	b_0'	.	.	b_0'	β	$-\alpha$	$-\gamma$	α	β	1
8	ω^3	1	$-\omega$	.	.	c_0'	.	.	c_0'	$-\gamma$	1	$-\alpha$	$-\beta$	α	1
9	$-\omega$	$-\omega^2$	1*	.	.	b_0	.	.	b_0'	$-\alpha$	β	1	$-\alpha$	β	1
10	1	ω^2	$-\omega^2$	k_{10}	.	.	k_{10}	.	.	1	$-\beta$	β	$-\beta$	-1	1
11	$-\omega^2$	ω	ω^2*	.	a_0	.	.	a_0'	.	β	α	$-\beta$	-1	γ	1
12	ω^2	0	ω	k_{12}	.	.	k_{12}	.	.	$-\beta$	0	α	$-\gamma$	0	1

if $\mathbf{w}^T\mathbf{G}^r(\omega, 1, 0)$, where $\mathbf{w}^T = [0, 0, 1]$, is an even power of ω. We find the same result from a corresponding analysis of the other columns of Table 3.5.1 so that:

THEOREM 5.6.1 *If in Φ a point $\mathbf{G}^r(\omega^s, 1, 0)$ lies on the line $z = \omega^t y$, then in Ψ the point $\mathbf{G}^r(\omega^s, 1, 0)$, interpreted in terms of miniquaternion elements from Table 5.6.2, lies on $z = \omega^t y$ or $z = \omega^{*t} y$ according as $\mathbf{w}^T\mathbf{G}^r(\omega^s, 1, 0)$ is an odd or an even power of ω.*

Next compare columns III and V in Table 5.6.3. We find that in Ψ A_r lies consistently on the various lines u_0 in the following way:

$$\begin{array}{cccc} a & a' & b' & c' \\ y = \gamma^* & y = \gamma & y = \beta & y = \alpha. \end{array}$$

This list can be completed by considering tables corresponding to Table 5.6.3 constructed for points B_r and C_r, so that we obtain:

THEOREM 5.6.2 *If, in Ψ, the point $U_0 = (\lambda, 1, 0)$, and*

$$\mathbf{G}^r(\lambda, 1, 0) = (\rho, \sigma, \tau),$$

then the point U_r lies on $y = \sigma\tau^{-1}$ or $y = (\sigma\tau^{-1})^*$ (calculated in $\mathscr{F}$).

The connecting link between Theorems 5.6.1 and 5.6.2 is to be found right back in Exercise 1.3.4: if $\sigma, \tau \in \{\pm\alpha, \pm\beta, \pm\gamma\}$ in either $\mathscr{F}$ or $\mathscr{Q}$ and multiplication is represented by $\times$ in $\mathscr{F}$ and $\otimes$ in $\mathscr{Q}$, then

$$\sigma \otimes \tau^{-1} = \begin{cases} \sigma \times \tau^{-1} & \text{when} \quad \tau \in \{\pm 1, \pm\beta\} = \{\pm 1, \pm\omega^2\}, \\ (\sigma \times \tau^{-1})^* & \text{when} \quad \tau \in \{\pm\alpha, \pm\gamma\} = \{\pm\omega, \pm\omega^3\}. \end{cases}$$

In effect, we have used this property in reverse and changed U_r in Φ to U'_r in Ψ when $\sigma \otimes \tau^{-1} = \sigma \times \tau^{-1}$, and preserved U_r as U_r when $\sigma \otimes \tau^{-1} = (\sigma \times \tau^{-1})^*$. This is of course only a matter of naming the points, and while it does not exactly match the algebraic specification, it provides a rather more convenient form for Table 5.4.4.

We have in fact devised another algebraic representation of Ψ, over the Galois field $\mathscr{F}$ rather than the miniquaternion system $\mathscr{Q}$, namely:

DEFINITION 5.6.1 Ψ is the set of points K_r, U_r, and lines $k_r, v_r, r \in \{0, \ldots, 12\}$, $U \in \{A, B, C, A', B', C'\}$, $v \in \{a, b, c, a', b', c'\}$, with incidence relations defined over the field $\mathscr{F}$ as follows:

(i) K_r, k_r form the real subplane.
(ii) $\mathbf{G}$ is a real Singer matrix.
(iii) $\{U_0\} = \{(\lambda, 1, 0)\}, \lambda \in \{\pm\omega, \pm\omega^2, \pm\omega^3\}$.
(iv) $U_r = \mathbf{G}^r U_0$.
(v) If $\mathbf{G}^r U_0 = (\rho, \sigma, \tau), r \in \{1, \ldots, 12\}$,

then
$$U_r \in k_r \quad \text{if} \quad \sigma\tau^{-1} \in \{0, \pm 1\},$$
$$U_r \in y = \sigma\tau^{-1} \quad \text{if} \quad \tau \in \{\pm\omega, \pm\omega^3\},$$
$$U_r \in y = (\sigma\tau^{-1})^* \quad \text{if} \quad \tau \in \{\pm 1, \pm\omega^2\}.$$

(vi) $\{v_0\} = \{y = \lambda\}, v_r = \mathscr{G}^r v_0$.
(vii) $U_r \in v_0 \Rightarrow U_{r+s} \in v_s$.

This is in effect the definition of Ψ originally devised by Veblen and Wedderburn in 1907: they viewed the plane as a hybrid between Φ and Ω. Their definition leads almost immediately to the combinatorial properties exhibited in Table 5.5.4, a table which greatly facilitates some investigations such as Theorem 5.5.2 and those which form the subject of the next section.

EXERCISE 5.6.1 Reconstruct Table 5.5.4 writing the columns in the order $ABC'CB'A'$ and the rows in the order $abc'cb'a'$, number the rows and columns 1 to 6 and rename the points K_r, $A_r, \ldots, A'_r$ as respectively $E_{0,r}, \ldots, E_{6,r}$ and the lines correspondingly as $e_{0,s}, \ldots, e_{6,s}$. The new table is Table 5.6.4.

Table 5.6.4

$$e_{0,s} = \{E_{i,s}, i \neq 0, E_{0,s}, E_{0,s+1}, E_{0,s+3}, E_{0,s+9}\}$$

$$e_{j,0} \supset E_{i,r} \Rightarrow e_{j,s} \supset E_{i,r+s}$$

	$A_s = E_{1,s}$	$B_s = E_{2,s}$	$C'_s = E_{3,s}$	$C_s = E_{4,s}$	$B'_s = E_{5,s}$	$A'_s = E_{6,s}$
$a_0 = e_{1,0}$	2, 6, 5	.	1, 8	.	9, 7	3, 11
$b_0 = e_{2,0}$	.	2, 6, 5	3, 11	.	1, 8	9, 7
$c'_0 = e_{3,0}$	1, 8	3, 11	2, 6, 5	9, 7	.	.
$c_0 = e_{4,0}$	.	.	9, 7	2, 6, 5	3, 11	1, 8
$b'_0 = e_{5,0}$	9, 7	1, 8	.	3, 11	2, 6, 5	.
$a'_0 = e_{6,0}$	3, 11	9, 7	.	1, 8	.	2, 6, 5

Prove that:

$$e_{i,0} \supset E_{j,s} \Rightarrow e_{3i,0} \supset E_{3j,3s} \quad (i, j \bmod 7;\ s \bmod 13),$$

so that $E_{j,s} \to E_{3j,3s}$ is a collineation; and that this collineation may be written as:

$$\mathscr{L}: A_r \to C'_{3r} \to B_{9r} \to A'_r \to C_{3r} \to B'_{9r} \to A_r, \quad K_r \to K_{3r}.$$

EXERCISE 5.6.2 The following table repeats Table 3.6.4, but the points are now to be regarded as points of Ψ. The points P_t are defined by $P_t = \mathbf{W}^t(1, 0, 0)$, where $\mathbf{W} = \omega\mathbf{1} + \mathbf{G}$, but the transformation represented by the matrix $\mathbf{W}$ is not a collineation in Ψ.

Table 5.6.5 $U_s \to P_t$

U_s	A_{6r}	B_{6r+7}	C'_{6r}	C_{6r-1}	B'_{6r+4}	A'_{6r+8}
P_t, $t =$	$7r+1$	$7r+2$	$7r+3$	$7r+4$	$7r+5$	$7r+6$

Construct the table derived from Table 5.5.4 by this substitution and verify that it has properties closely resembling those of Table 3.6.5.

EXERCISE 5.6.3 It can be observed that Table 5.6.4 is symmetrical about both diagonals. That is, for symmetry about the leading diagonal:

$$E_{i,s} \in e_{j,0} \Rightarrow E_{j,s} \in e_{i,0}.$$

(i) Prove that

$$E_{i,r} \in e_{j,t} \Rightarrow E_{j,r} \in e_{i,t} \Rightarrow E_{j,-t} \in e_{i,-r},$$

and hence that

$$\mathscr{J}': E_{i,r} \rightarrow e_{i,-r} \text{ is a polarity.}$$

(ii) For the other diagonal: $E_{i,s} \in e_{j,0} \Rightarrow E_{-j,s} \in e_{-i,0}$. Prove that this relation generates the polarity

$$\mathscr{J}^{*\prime}: E_{i,r} \rightarrow e_{-i,-r}.$$

(iii) Prove that in each of the two polarities the real absolute points are K_0, K_7, K_8, K_{11} and that these points constitute the real conic $y^2+z^2-xz = 0$.

EXERCISE 5.6.4 For the polarity $\mathscr{J}^{*\prime}$ verify that:

(i) $K_r \rightarrow k_{-r}$, $U_r \rightarrow u'_{-r}$ (that is, $A_r \rightarrow a'_{-r}$ etc.).

(ii) The 16 absolute points are K_0, K_7, A_8, A_{12} and the points obtained from these by the operations of the cyclic group generated by the operation $\mathscr{L}$ defined in Exercise 5.6.1.

(iii) Of the lines in the plane 12 contain four absolute points each, 48 contain two each, 16 contain one each and 15 contain no absolute points. Three of these 15 lines are the sides of the diagonal triangle $K_4K_{10}K_{12}$ of $K_0K_7K_8K_{11}$, and the remaining 12 pass four through each of the vertices of this triangle.

(iv) Construct the table corresponding to Table 5.4.1.

EXERCISE 5.6.5 *An oval in* Ψ'. (i) Prove that under the non-trivial homology $\mathscr{H}_{9,4}$ in Δ_0 with centre K_9 and axis k_4, K_0 and K_7 are fixed and K_8 and K_{11} are interchanged. Using Table 5.6.4 determine the set of six points such as $\mathscr{H}_{9,4}A_1$ which are obtained from A_1 by applying the homologies (in the group of projectivities ∇) which fix two of the points K_0, K_7, K_8, K_{11} and interchange the other two points.

(ii) Prove that the set of six points so obtained is left invariant by every collineation in ∇ which permutes the points K_0, K_7, K_8, K_{11}, and by the collineation

$$\mathscr{A}: K_r \to K_r, \quad A_r \leftrightarrow C'_r, \quad B_r \leftrightarrow A'_r, \quad C_r \leftrightarrow B'_r.$$

(iii) By considering typical pairs of points in relation to the group of 48 collineations defined in part (ii), prove that the set of ten points consisting of K_0, K_7, K_8, K_{11} and the six points derived from A_1 in part (i) form an oval, say $\mathscr{O}_3$.

In §4.6 we constructed an oval $\mathscr{O}_1$, in the plane Ω, which does not determine a polarity (Ω not being self-dual). We shall find that $\mathscr{O}_3$ does define a polarity, and that this polarity $\mathscr{J}'$ (which is of the same type as $\mathscr{J}$) is an extension to Ψ of the polarity determined by the conic $\{K_0, K_7, K_8, K_{11}\}$, in Δ_0. In Table 5.6.6 the entries are the following points and lines:

Borders: vertices and tangents of $\mathscr{O}_3$; e.g. c_{10} is the tangent at C_3.

Lower triangle: exterior points; e.g. $B'_8 = c_{10} \cap a_{12}$.

Upper triangle: chords; e.g. $b'_5 = C_3 A_1$.

From Table 5.6.6 we see that the tangent at a vertex U_r or K_r of $\mathscr{O}_3$ is in every case the line u_{-r} or k_{-r} (respectively). Now the correspondence $U_r \leftrightarrow u_{-r}$, $K_r \leftrightarrow k_{-r}$ (for all U and all r) determines the polarity $\mathscr{J}'$ (Exercise 5.6.3). The set of absolute points in $\mathscr{J}'$ comprises K_0, K_7, K_8, K_{11} and the 18 points

$$U_r, \; r \in \{\tfrac{1}{2}.2, \tfrac{1}{2}.6, \tfrac{1}{2}.5\} = \{1, 3, 9\}, \quad U \in \{A, B, C, A', B', C'\}.$$

That is, the points are the vertices of the regular quadrangle $K_0 K_7 K_8 K_{11}$ and the six complex points on each side of the diagonal triangle. The oval $\mathscr{O}_3$ consists of the quadrangle together with a pair of points from each of the sets of six. Since every point of the plane lies on at least two chords of the oval, the polarity is completely determined by its effect on the vertices of $\mathscr{O}_3$ (namely sending them to the corresponding tangents). The set of absolute points in $\mathscr{J}'$ in fact consists of three ovals, any of which

Table 5.6.6 *Vertices, tangents, chords and exterior points of the oval $\mathcal{O}_3$*

	K_0	K_7	K_8	K_{11}	A_1	C_3	B_9	C'_1	B'_3	A'_9	
k_0		k_4	k_{12}	k_{10}	c'_0	b'_0	a'_0	a_0	c_0	b_0	K_0
k_6	K_9		k_7	k_{11}	b'_7	c'_7	b_7	c_7	a_7	a'_7	K_7
k_5	K_1	K_6		k_8	a_8	a'_8	b'_8	c'_8	b_8	c_8	K_8
k_2	K_3	K_2	K_5		a'_{11}	c_{11}	c'_{11}	b_{11}	b'_{11}	a_{11}	K_{11}
a_{12}	C'_0	B'_6	A_5	A'_2		b'_5	c'_6	k_1	a_9	a'_3	A_1
c_{10}	B'_0	C'_6	A'_5	C_2	B'_8		a'_2	c'_9	k_3	c_1	C_3
b_4	A'_0	B_6	B'_5	C'_2	C'_7	A'_{11}		b_3	b'_1	k_9	B_9
c'_{12}	A_0	C_6	C'_5	B_2	K_{12}	C'_4	B_{10}		c_5	a_6	C'_1
b'_{10}	C_0	A_6	B_5	B'_2	A_4	K_{10}	B'_{12}	C_8		b_2	B'_3
a'_4	B_0	A'_6	C_5	A_2	A'_{10}	C_{12}	K_4	A_7	B_{11}		A'_9
	k_0	k_6	k_5	k_2	a_{12}	c_{10}	b_4	c'_{12}	b'_{10}	a'_4	

determines the polarity, the other two being derived from $\mathcal{O}_3$ by the collineation

$$\mathscr{C}\mathscr{A}: (K_r, A_r, B_r, C_r, A'_r, B'_r, C'_r) \to (K_r, B_r, C_r, A_r, C'_r, A'_r, B'_r)$$

and its square.

EXERCISE 5.6.6 Verify that the set of points

$$\{(\rho, j_\rho \rho^*, 1) : \rho \in \mathscr{Q}, \rho \neq 0\} \cup \{(1,0,0), (0,1,0)\}$$

forms an oval. (If $\rho = r + \beta r'$, then $j_\rho = r^2 + r'^2$ and $\rho^* = r - \beta r'$.)

5.7 The subplanes of Ψ

In § 5.3 we discussed two configurations constructed from a regular quadrangle with three real vertices (Diagram 5.3, Theorems 5.3.2, 5.3.3). A rather more incisive statement of the theorems can be made if we express them in terms of the operation of 'completing a quadrangle' (Definition 2.1.6).

THEOREM 5.7.1 *In* Ψ *the completion of a regular quadrangle* $XYZK$ *with three real vertices* X, Y, Z *is* (i) *a subplane* Δ *of order* 3 *if at least one of the lines* KX, KY, KZ *is real*; (ii) *a subplane of order* 2 *if none of these lines is real.*

Later in this section examples will be given of other subplanes in Ψ, but first we enumerate the subplanes of the two types described in Theorem 5.7.1. Taking first the subplanes of order 3, we see that if the complete configuration represented in Diagram 5.3 is a subplane of order 3 (other than the real subplane Δ_0), then it can be determined by the following data: the set of four real collinear points $\{X, Y, N, N'\}$, one other real point, Z, and a complex point (K) on one of the lines ZX, ZY, ZN, ZN'. The gross number of such selections of points in Ψ is given by:

choice of a real line: 13,
choice of a non-incident real point: 9,
choice of the complex point, K: 4×6.

Within the plane Δ, when it has been constructed, there is no choice of real line or disjoint real point, while the number of choices of complex point is 4×2. Thus

THEOREM 5.7.2 *The number of subplanes* Δ *in* Ψ *constructible as in Theorem* 5.7.1 *is* $\frac{1}{8}.13.9.24 = 351$.

A Fano subplane is uniquely determined by three non-collinear real points, X, Y, Z say, and a point F such that none of FX, FY, FZ is real. In Theorem 5.3.4 we proved that there are 24 points with this property in relation to X, Y, Z. In the Fano subplane X, Y, Z are the only real points, and F is the only point which forms with them a regular quadrangle, so that each choice of X, Y, Z (in any order) and F provides a unique set of points in the subplane that the points determine. The number of choices in Δ_0 of X, Y, Z is $\frac{1}{6}.13.12.9 = 234$, thus

THEOREM 5.7.3 *The number of Fano planes in* Ψ *of the type that contains three real points is* $234.24 = 5616$.

5616 is the order of the group of projectivities ∇, and we prove next

THEOREM 5.7.4 *The group of projectivities ∇ is transitive on the set of Fano subplanes of the type defined in Theorem* 5.7.3.

First we can map any three non-collinear real points to X, Y, Z and have therefore only to prove that we can map the Fano subplane determined by X, Y, Z, G to that determined by X, Y, Z, F for any choice of two points in the set of 24 which satisfy the required conditions. Take $F = (\rho, \sigma, 1)$, $G = (\lambda, \mu, 1)$, where $\rho, \sigma, \lambda, \mu \in \mathscr{Q}^*$ and $\rho\sigma \neq \pm 1$ and $\lambda\mu \neq \pm 1$. We have to show that there is a projectivity that maps X, Y, Z to X, Y, Z in any order, and G to F. The group of projectivities satisfying the conditions in relation to X, Y, Z is compounded of the sign-changing group $(a, b, 1) \rightarrow (\pm a, \pm b, 1)$ of order 4 and the permutation group on the coordinates of order 6. We have only to show that the 24 transforms of $(\lambda, \mu, 1)$ are all different, since then exactly one of them is $(\rho, \sigma, 1)$. The transforms are:

$$(\pm\lambda, \pm\mu, 1) \quad \text{and} \quad (\pm\mu, \pm\lambda, 1);$$

$$(\pm\lambda, \pm 1, \mu) = (\pm\nu, \pm\mu, 1) \quad \text{and} \quad (\pm\mu, \pm\nu, 1);$$

$$(\pm 1, \pm\mu, \lambda) = (\pm\lambda, \pm\nu, 1) \quad \text{and} \quad (\pm\nu, \pm\lambda, 1),$$

where $\nu = \lambda\mu$. Under the prescribed conditions, namely $\lambda, \mu \in \mathscr{Q}^*$ and $\lambda\mu \neq \pm 1$, we have $\{\pm\lambda, \pm\mu, \pm\nu\} = \{\pm\alpha, \pm\beta, \pm\gamma\}$, and consequently all 24 points listed above are different. The theorem is therefore valid. The other group Σ of collineations of Ψ has no further effect, since the set of points $\{\mathscr{A}_\sigma(\lambda, \mu, 1) : \mathscr{A}_\sigma \in \Sigma\}$ is a subset of the above set of 24.

We conclude by listing the other subplanes of Ψ. In a recent paper R. H. F. Denniston (1968) constructs two other types of subplanes of order 3 and two of order 2, and finds that, since the total numbers of subplanes of the two orders that he has constructed are equal to the totals found (without classification) by a computer, there can be no other types of subplanes in Ψ. To simplify comparison with Denniston's account we print, in the reference to his paper in the list of References, a table setting out the correspondence between Denniston's notation and that used here.

SUBPLANES OF ORDER 3

Type I: 4 collinear real points and one other real point. 4 real lines concurrent in the single point, one real line containing the collinear real points (Theorem 5.7.1)

K_0	K_1	K_5	K_3	A_5	A_2	A_9	B'_9	B_4	K_9	B'_5	B'_2	C'_4
K_1	K_5	K_3	A_5	A_2	A_9	B'_9	B_4	K_9	B'_5	B'_2	C'_4	K_0
K_3	A_5	A_2	A_9	B'_9	B_4	K_9	B'_5	B'_2	C'_4	K_0	K_1	K_5
K_9	B'_5	B'_2	C'_4	K_0	K_1	K_5	K_3	A_5	A_2	A_9	B'_9	B_4
k_0	k_5	k_2	a_3	a_0	c'_1	k_9	b'_3	b'_9	a_9	b'_0	b_1	k_4

Number of planes: 351 = 5616/16.

Type II: 4 collinear real points, 4 real lines through one of these points

K_0	K_1	A_{10}	K_3	C_4	B_4	B_{12}	A_4	C_{10}	K_9	C_{12}	A_{12}	B_{10}
K_1	A_{10}	K_3	C_4	B_4	B_{12}	A_4	C_{10}	K_9	C_{12}	A_{12}	B_{10}	K_0
K_3	C_4	B_4	B_{12}	A_4	C_{10}	K_9	C_{12}	A_{12}	B_{10}	K_0	K_1	A_{10}
K_9	C_{12}	A_{12}	B_{10}	K_0	K_1	A_{10}	K_3	C_4	B_4	B_{12}	A_4	C_{10}
k_0	b'_1	b'_3	a'_3	k_4	c'_1	c'_9	c'_3	a'_9	b'_9	k_{12}	a'_1	k_{10}

Number of planes: 104 = 5616/54.

Type III: 1 real point, 1 real line through it

K_0	A_0	A_1	B_0	A'_6	C_8	A_7	B_9	C'_5	C_0	B'_2	C_3	B_{11}
A_0	A_1	B_0	A'_6	C_8	A_7	B_9	C'_5	C_0	B'_2	C_3	B_{11}	K_0
B_0	A'_6	C_8	A_7	B_9	C'_5	C_0	B'_2	C_3	B_{11}	K_0	A_0	A_1
C_0	B'_2	C_3	B_{11}	K_0	A_0	A_1	B_0	A'_6	C_8	A_7	B_9	C'_5
k_0	a_8	b'_5	a'_4	a'_0	c'_{12}	c'_6	b_7	c_{11}	b'_{10}	b'_0	a'_2	c'_0

Number of planes: 624 = 5616/9.

Total number of planes of order 3 (including Δ_0): 1080, quite surprisingly the same as the number of such planes in Ω (Exercise 4.5.2), and one-seventh of the number in Φ (Exercise 3.4.1).

SUBPLANES OF ORDER 2

Type I: 3 real points, 3 real lines joining them by pairs

K_0	K_1	K_2	B_0	C_1	A_3	A'_{12}
K_1	K_2	B_0	C_1	A_3	A'_{12}	K_0
B_0	C_1	A_3	A'_{12}	K_0	K_1	K_2
k_0	k_1	c'_2	a'_6	a'_0	a_1	k_{12}

Number of planes: 5616 (Theorem 5.7.3).

Type II: One real point, one disjoint real line

K_0	A_2	A'_2	A_5	C'_2	A_1	A_{11}
A_2	A'_2	A_5	C'_2	A_1	A_{11}	K_0
A_5	C'_2	A_1	A_{11}	K_0	A_2	A'_2
a_0	k_2	a_{12}	c'_{10}	c'_0	a_9	a'_0

Number of planes: 5616×3.

Type III: No real point, no real line

A_2	A_6	A'_4	A_5	A'_{12}	A'_7	A_{10}
A_6	A'_4	A_5	A'_{12}	A'_7	A_{10}	A_2
A_5	A'_{12}	A'_7	A_{10}	A_2	A_6	A'_4
a_0	a_1	a'_2	a'_7	a_9	a_4	a'_{12}

Number of planes: 5616×2.

Total number of planes of order 2: 5616×6. (Compare the quite different number in Ω, namely 51 840 (which contains, for example, no factor 13)—Exercise 4.4.3.)

EXERCISE 5.7.1 *Intersections of pairs of subplanes of order* 3. Since two distinct planes of order 3 cannot have common a regular quadrangle, so far as points only are concerned, the possible intersections of two such planes are:

(0) the planes are disjoint, (i) one point, (ii) two points;
(iii) three points, (*a*) collinear, (*b*) not collinear;
(iv) four points, (*a*) collinear, (*b*) only three collinear;
(v) five points, four collinear.

In Φ we found all the above types of intersection except (iii*a*) and (iv*b*). In Ψ the intersection of Δ_0 with any other subplane can be found immediately from the tables above, and in particular we find that no subplane of order 3 is disjoint from Δ_0. The possible intersection-structures in Δ_0 are in fact (v), (iv$a$) and (i).

To determine the possible intersections for pairs of subplanes not involving Δ_0 we fix one of the planes and investigate its relations to typical members of the set obtained from another subplane by applying operations of ∇ and Σ. We find easily examples (besides those involving Δ_0) of all types of incidence-structure except (0) and (iii$a$).

Question I: Is there any pair of disjoint subplanes of order 3 in Ψ? II: Is there a pair for which the intersection is three collinear points? III: If the pair has a single common point, does it have a single common line and are the two incident? IV: Are the numbers of common points and lines always equal?

LIST OF SPECIAL SYMBOLS

Algebraic

$\mathscr{D}$	the Galois field of order 3, subset of $\mathscr{F}$ and $\mathscr{Q}$ (§ 1.1)
$\mathscr{F}$	the Galois field of order 9 (§ 1.2)
$\mathscr{Q}$	the miniquaternion system $\{0, \pm 1, \pm\alpha, \pm\beta, \pm\gamma\}$ (§ 1.3)
$\mathscr{Q}^*$	$\mathscr{Q}^* = \mathscr{Q} - \mathscr{D} = \{\pm\alpha, \pm\beta, \pm\gamma\}$ (§ 1.3)
Aut ($\mathscr{Q}$)	the automorphism group of $\mathscr{Q}$ (§ 1.4)
$\mathfrak{K}$	an arbitrary (not necessarily commutative) field
$\mathscr{K}$	an arbitrary commutative field
ϵ	$\epsilon = (0, 1)$, $\epsilon^2 = -1$; element of $\mathscr{F}$ (§ 1.2)
ω	$\omega = 1 - \epsilon$, $\omega^8 = 1$; element of either $\mathscr{F}$ or $\mathscr{Q}$ (§1.2, §1.3)
α	$\alpha = \omega$, $\alpha \otimes \alpha = -1$; element of $\mathscr{Q}$ (§ 1.3)
β	$\beta = \omega^6$, $\beta \otimes \beta = -1$; element of $\mathscr{Q}$ (§ 1.3)
γ	$\gamma = \omega^7$, $\gamma \otimes \gamma = -1$; element of $\mathscr{Q}$ (§ 1.3)
$\mathscr{E}$	even powers of ω (§ 1.3)
$\mathscr{O}$	odd powers of ω (§ 1.3)

Planes

$\Pi(\mathfrak{K})$	the plane over $\mathfrak{K}$ (§ 2.3)
Δ	$\Delta = \Pi(\mathscr{D})$, the plane of order 3 (§ 3.1)
Φ	$\Phi = \Pi(\mathscr{F})$, the Galois plane of order 9 (§ 3.3)
Ω	translation plane of order 9 (§ 4.1)
Ω^D	dual translation plane of order 9 (§ 4.7)
Ψ	Hughes plane of order 9 (§ 5.1)

Collineations, Correlations

G	a Singer matrix for Δ (§ 3.2)
S	a Singer matrix for Φ (§ 3.6)

$\mathscr{J}$ (i) the conjugation collineation of Φ (§ 3.3);
(ii) a polarity of Ψ (§ 5.2, § 5.4)

$\mathscr{J}^*$, $\mathscr{J}'$, $\mathscr{J}^{*\prime}$ polarities of Ψ (§ 5.4, § 5.6)

$\mathscr{R}$, $\mathscr{S}_{\rho,\sigma}$, $\mathscr{T}_{\delta,\epsilon}$, $\mathscr{U}$ collineations of Ω (§ 4.2)

$\mathscr{A}_\lambda$ collineations of Ω, or of Ψ (§ 4.2, § 5.1)

$\mathfrak{G}$ full collineation group of Ω (§ 4.3)

∇ projective group of Ψ (§ 5.1)

Σ $\Sigma = \{\mathscr{A}_\lambda : \lambda \in \mathscr{Q}^*\}$, a collineation group in Ψ (§ 5.1)

REFERENCES

André, J. (1955). Projektive Ebenen über Fastkörpern. *Math. Z.* **62**, 137–60.

Denniston, R. H. F. (1968).† Subplanes of the Hughes plane of order 9. *Proc. Camb. Phil. Soc. math. phys. Sci.* **64**, 589–98.

Dickson, L. E. (1905). On finite algebras. *Nach. Ges. Wiss Göttingen*, pp. 358–93.

Hall, M. (1959). *The Theory of Groups*. New York: Macmillan.

Herstein, I. N. (1964). *Topics in Algebra*. New York, Toronto, London: Blaisdell.

Hughes, D. R. (1957). A class of non-Desarguesian projective planes. *Can. J. Math.* **9**, 378–88.

Ostrom, T. G. (1964). Finite planes with a single (p, L)-transitivity. *Arch. Math.* **15**, 378–84.

Rodriguez, G. (1959). Un esempio di ovale che non è quasi-conica. *Boll. Un. math. ital* (3), **14**, 500–3.

Room, T. G. (1967). *A Background to Geometry*. Cambridge University Press.

Segre, B. (1955). Ovals in a finite projective plane. *Can. J. Math.* **7**, 414–16.

Singer, J. (1938). A theorem in finite projective geometry and some applications to number theory. *Trans. Am. Math. Soc.* **43**, 377–85.

Veblen, O. and Wedderburn, J. M. (1907). Non-Desarguesian and non-Pascalian geometries. *Trans. Am. Math. Soc.* **8**, 379–88.

Zappa, G. (1957). Sui gruppi di collineazioni dei paini di Hughes. *Boll. Un. mat. ital.* (3), **12**, 507–16.

Zassenhaus, H. (1936). Über endliche Fastkörper. *Abh. Hamburg* **11**, 187–220.

† Our notation for the points of Ψ corresponds to Denniston's as follows:

D	A	B	C	D	E	F	G	H	I	J	K	L	M
R & K	K_0	K_1	K_2	K_3	K_4	K_5	K_6	K_7	K_8	K_9	K_{10}	K_{11}	K_{12}
D main symbol	N	O	P	Q	R	S	T	U	V	W	X	Y	Z
R & K subscript	0	1	2	3	4	5	6	7	8	9	10	11	12
D subscript	3	4	5	6	7	8							
R & K main symbol	A	C	B	B'	A'	C'							

SUGGESTIONS FOR FURTHER READING

For the most complete bibliography up to 1968 on the subject of projective planes, see the book by P. Dembowski (1968) below.

BOOKS

Albert, A. A. and Sandler, R. (1968). *An Introduction to Finite Projective Planes.* New York: Holt, Rinehart and Winston.

Blattner, J. W. (1968). *Projective Plane Geometry.* San Francisco: Holden-Day.

Blumenthal, L. M. (1961). A *Modern View of Geometry.* San Francisco, London: W. H. Freeman and Company.

Bumcrot, R. J. (1969). *Modern Projective Geometry.* New York: Holt, Rinehart and Winston.

Dembowski, P. (1968). *Finite Geometries. Ergebnisse der Mathematik und ihrer Grenzgebiete,* Volume 44. New York: Springer Verlag.

Pickert, G. (1955). *Projektive Ebenen.* Berlin, Göttingen, Heidelberg: Springer.

PLANES OF ORDER NINE AND RELATED PLANES

André, J. (1954). Über nicht-Desarguessche Ebenen mit transitiver Translationsgruppe. *Math. Z.* **60**, 156–86.

Bose, R. C. and Chakravarti, I. M. (1965). Hermitian varieties in a finite projective space $PG(N, q^2)$. *Lecture notes.* Chapel Hill, N.C.

Corsi, G. (1959). Sui triangoli omologici in un piano sul quasicorpo associativo di ordine 9. *Matematiche* **14**, 40–66.

Hall, M., Swift, J. D. and Killgrove, R. (1959). On projective planes of order nine. *Math. Comp.* **13**, 233–46.

Magari, R. (1958). Le configurazioni parziali chiuse contenute nel piano, *P*, sul quasicorpo associativo di ordine 9. *Boll. Un. mat. ital.* **13**, 128–40.

Parker, E. T. and Killgrove, R. B. (1964). A note on projective planes of order nine. *Math. Comp.* **18**, 506–8.

Pickert, G. (1965). Zur affinen Einführung der Hughes-Ebenen. *Math. Z.* **89**, 199–205.

Ostrom, T. G. (1965). A characterization of the Hughes planes. *Canad. J. Math.* **17**, 916–22.

Rodriguez, G. (1964). Le sub-collineazioni nei piani di translazione sopra i quasicorpi associativi. *Matematiche* (*Catania*) **19**, 11–18.

Room, T. G. (1969). Geometry in a class of near-field planes. *J. Lond. Math. Soc.* (2), **1**, 591–605.

Room, T. G. (1969). Veblen-Wedderburn hybrid planes. *Proc. Roy. Soc. Lond.* A **309**, 157–70.

Room, T. G. (1970). The combinatorial structure of the Hughes plane. *Proc. Camb. Phil. Soc.*, to appear.

Room, T. G. (1971). Polarities and ovals in the Hughes plane. *J. Austral. Math. Soc.*, to appear.

Rosati, L. A. (1958). I gruppe di collineazioni dei piani di Hughes. *Boll. Un. mat. ital.* **13**, 505–13.

Rosati, L. A. (1960). Unicità e autodualità dei piani di Hughes. *Rc. Semin. math. Univ. Padova* **30**, 316–27.

COLLINEATIONS AND CORRELATIONS

André, J. (1958). Über Perspektivitaten in endlichen affinen Ebenen. *Arch. Math.* **9**, 228–35.

Baer, R. (1946). Polarities in finite projective planes. *Bull. Amer. Math. Soc.* **52**, 77–93.

Cofman, J. (1965). Homologies of finite projective planes. *Arch. Math.* **16**, 476–79.

Hoffman, A. J., Newman, M., Strauss, E. G. and Taussky, O. (1956). On the number of absolute points of a correlation. *Pac. J. Math.* **6**, 83–96.

Ostrom, T. G. and Wagner, A. (1959). On projective and affine planes with transitive collineation groups. *Math. Z.* **71**, 186–99.

Piper, F. C. (1965). Collineation groups containing elations. *Math. Z.* **89**, 181–91.

Piper, F. C. (1967). Collineation groups containing homologies. *J. Algebra* **6**, 256–69.

Wagner, A. (1959). On perspectivities of finite projective planes. *Math. Z.* **71**, 113–23.

OVALS

Bartocci, U. (1967). Una nuova classe di ovali proiettive finite. (English summary.) *Atti. Accad. naz. Lincei Rc.* (Cl. Sci. Fis. Mat. Nat.) (8), **43**, 312–16.

Buekenhout, F. (1966). Etude intrinsèque des ovales. *Rc. Mat. Applic.* **25**, 1–61.

Ostrom, T. G. (1955). Ovals, dualities, and Desargues's theorem. *Canad. J. Math.* **7**, 417–31.

Quist, B. (1952). Some remarks concerning curves of the second degree in a finite plane. *Ann. Acad. Sci. Fenn.* no. 134, 1–27.

CONSTRUCTION AND UNIQUENESS OF PLANES

Hall, M. (1953). Uniqueness of the projective plane with 57 points. *Proc. Amer. Math. Soc.* **4**, 912–16. (Correction, same journal (1954), **5**, 994–7.)

Hall, M., Swift, J. D. and Walker, R. J. (1956). Uniqueness of the projective plane of order eight. *Mathl Tabl. natn. Res. Coun., Wash.* **10**, 186–94.

Ostrom, T. G. (1968). Vector spaces and construction of finite projective planes. *Arch. Math.* **19**, 1–25.

Tarry, G. (1901). Le problème des 36 officiers. *C. R. Ass. fr. Avanc. Sci. Nat.* **2**, 170–203.

INDEX